**R. K. Doharey**
**Aman Verma**
**Satyapriya**
**Amrit Warshini**
Editors

# Global Horizons in Agricultural Extension Education

## Insights from International Perspectives

nova
science publishers
www.novapublishers.com

DOI: https://doi.org/10.52305/RQLT7136



Additional color graphics may be available in the e-book version of this book.

**Library of Congress Cataloging-in-Publication Data**

ISBN: 979-8-89530-446-4 (Hardcover)
ISBN: 979-8-89530-501-0 (eBook)

*Published by Nova Science Publishers, Inc. † New York*

*To the farmers worldwide, the heartbeat of our global food systems, whose resilience, wisdom, and deep connection to the land inspire us daily. Your dedication to sustaining life on this planet through your skilled hands and profound knowledge is a testament to humanity's perseverance and our sacred bond with nature.*

*To the extension workers, educators, and changemakers tirelessly working on the frontlines, bridging the gap between research and practice. Your commitment to empowering farming communities, promoting sustainable agriculture, and fostering inclusive development uplifts us all.*

*To the researchers, academics, and thought leaders who continually push the boundaries of knowledge and challenge us to reimagine agricultural extension. Your insights, innovations, and unwavering pursuit of truth illuminate the path toward a more equitable and sustainable future.*

*To the policymakers and global leaders championing the transformative power of agricultural extension. Your vision and determination to create an enabling environment for positive change are vital catalysts for progress.*

*And to the youth, the torchbearers of tomorrow's agriculture, whose curiosity, passion, and innovative spirit will shape the horizons of extension education for generations to come. May this work inspire you to embrace the challenges and opportunities that lie ahead with unwavering optimism and steadfast resolve.*

*This book is a celebration of the collective wisdom, diverse perspectives, and collaborative efforts that propel agricultural extension into new frontiers, transcending boundaries and uplifting communities worldwide. May it serve as a beacon of hope and a call to action, reminding us all of our shared responsibility to nourish our planet, save trees, save water, save farmers, and save the earth.*

*R. K. Doharey*
*Aman Verma*
*Satyapriya*
*Amrit Warshini*
Editors

# Contents

# Forewords

## Udham Singh Gautam

Deputy Director General (Agril. Extn.)

I am delighted to write this foreword for *"Global Horizons in Agricultural Extension Education: Insights from International Perspectives."* The field of agricultural extension has been pivotal in driving sustainable agricultural development, enhancing food security, and improving the livelihoods of farming communities across the globe.

As the world grapples with the challenges of climate change, population growth, and diminishing natural resources, the role of agricultural extension has become increasingly crucial. Extension professionals act as vital conduits between scientific research and practical application, facilitating the transfer of knowledge and innovative practices to farmers. This empowers them to adopt sustainable and climate-resilient agricultural methods, ensuring the longevity and health of our agricultural systems.

As we look ahead, the future of agricultural extension lies in its ability to adapt to changing global dynamics, leverage emerging technologies, and foster collaborative partnerships. This is a timely publication for practitioners, students, or policymakers as it offers valuable insights and guidance in the field of agricultural extension.

I congratulate the editors and authors for their dedication and commitment to advancing the field of agricultural extension education by bringing out this publication. It is a significant contribution to the field and will undoubtedly inspire extension professionals.

**R. S. Kureel**
Vice-Chancellor
Mahatma Gandhi University of Horticulture and Forestry
Durg, Chhattisgarh, India

The agricultural sector, especially horticulture and forestry sector is facing a critical juncture with unprecedented challenges and opportunities. "Global Horizons in Agricultural Extension Education" explores cutting-edge trends and best practices shaping the future of extension education worldwide.

Mahatma Gandhi University of Horticulture and Forestry is committed to advancing extension services for the betterment of stakeholders especially post harvest Management and processing technology, value addition and products development. This book aligns perfectly with our mission, offering valuable insights and strategies for enhancing program effectiveness in horticulture and forestry.

The authors skillfully curate diverse perspectives, delving into digital tools (Al), empowering rural women, cultivating climate resilience, and fostering cross-cultural collaboration. It sheds light on the complexities extension professionals face while offering solutions to overcome them.

The book emphasizes the importance of integrating local realities with global agendas in today's interconnected world. It challenges us to reimaging extension education for a more inclusive, equitable, and sustainable approach.

I commend the authors for this comprehensive resource, an essential read for extension professionals, policymakers, researchers, and other stakeholders committed to advancing sustainable agriculture in India and worldwide. By embracing the presented strategies, we can collectively strengthen extension education systems and input support services to empower farming community and other stakeholders, paving the way for more resilient and prosperous future for agriculture sector globally.

## S. V. Suresha

University of Agricultural Sciences, Bangalore
Vice-Chancellor

Agricultural extension, a specialized field within agriculture, plays a pivotal role in transferring diverse aspects of technology to farmers. Its significance lies in bridging the gap between research and practical farming, facilitating the dissemination of crucial knowledge and practices essential for sustainable agricultural development worldwide. This field ensures that scientific advancements effectively reach farmers, thereby enhancing agricultural productivity and sustainability in tangible ways.

*"Global Horizons in Agricultural Extension Education: Insights from International Perspectives"* offers a comprehensive exploration of this vital sector. The book meticulously covers contemporary issues, global trends, case studies, and practical applications. It stands as an indispensable resource for educators, researchers, policymakers, and practitioners dedicated to advancing agricultural education and communication.

This book emphasizes the importance of equipping farmers with the knowledge and tools they need to adapt and thrive. It highlights innovative approaches and the integration of technology to enhance extension services, such as digital extension tools that transform information sharing and utilization. Additionally, it emphasizes the necessity of gender-responsive agricultural extension to empower rural women by ensuring equal access to resources and opportunities.

The adoption of climate-resilient technologies and practices is crucial for building resilience against climate change. The book provides detailed insights into precision farming, supported by advancements in artificial intelligence and machine learning, enabling farmers to make data-driven decisions that optimize resource use and improve crop yields. It also explores the potential of secondary agriculture and sustainable food value chains, advocating for a holistic approach to agricultural development.

The adoption of these modern technologies and methodologies will bring significant advancements in agricultural extension, making the insights and approaches detailed in this book immensely valuable to extension professionals.

I commend the authors for their rigorous research and valuable insights by emphasizing the transformative power of education in improving agricultural practices and livelihoods. As we navigate the complexities of the

$21^{st}$ century, "Global Horizons in Agricultural Extension Education" will be a pivotal guide in our journey towards a more sustainable and prosperous agricultural future.

I am glad to provide this foreword and wholeheartedly endorse this book to all individuals engaged in agriculture. I extend my sincere congratulations to the team for producing this valuable and informative publication.

# Preface

In an increasingly globalized world, the field of agricultural extension is undergoing profound transformations. Traditional extension models are challenged to adapt to rapidly evolving socio-economic, technological, and environmental landscapes. As we grapple with the pressing challenges of climate change, food insecurity, and rural poverty, it is imperative that we rethink and reimagine extension approaches to meet the diverse needs of farmers and rural communities across the globe.

This book, "Global Horizons in Agricultural Extension Education: Insights from International Perspectives," offers a comprehensive exploration of the latest trends, innovations, and best practices in agricultural extension from a global standpoint. Drawing upon the collective wisdom of experts from various corners of the world, this volume provides a rich tapestry of insights and experiences that shed light on the complexities and opportunities within the field.

The chapters in this book delve into a wide range of topics, from unpacking global agendas and local realities to embracing technological disruptions and leveraging digital tools for Extension 2.0. We explore the art of building bridges through intercultural communication and collaboration, empowering rural women through gender-responsive extension, and cultivating resilience in the face of climate change.

Furthermore, this book examines the critical issue of evaluating impact and measuring success from a global perspective, offering valuable insights into the future of extension and its role in promoting sustainable agriculture. It also explores emerging trends such as gamification, agroecology, and integrating artificial intelligence and big data in extension services.

This book aims to serve as a catalyst for dialogue, collaboration, and innovation within the global extension community by bringing together diverse voices and perspectives. We hope that the insights and ideas presented in these pages will inspire practitioners, policymakers, and researchers alike to embrace a more inclusive, sustainable, and impactful approach to agricultural extension education.

As we navigate the complexities of the $21^{st}$ century, it is clear that the challenges facing agriculture and rural development are intrinsically global in nature. Only through a shared understanding and a willingness to learn from one another can we truly unlock the full potential of agricultural extension to create a more food-secure, equitable, and resilient world.

*R. K. Doharey*
*Aman Verma*
*Satyapriya*
*Amrit Warshini*
Editors

# Acknowledgments

Creating *"Global Horizons in Agricultural Extension Education: Insights from International Perspectives"* has been a collaborative effort, and we are deeply grateful to all who contributed to this endeavor.

Firstly, we extend our heartfelt thanks to all the contributing authors whose dedication and expertise have enriched this book. Your insights and research have provided a comprehensive understanding of the various dimensions of agricultural extension education in a global context.

We express our profound gratitude to Acharya Narendra Deva University of Agriculture & Technology, Ayodhya, U.P., India, and ICAR-Indian Agricultural Research Institute, New Delhi, for their continuous support and encouragement. The resources and opportunities provided by these esteemed institutions have been instrumental in bringing this book to fruition.

We are also thankful to the editorial team for their meticulous efforts in reviewing and refining the manuscript. Your attention to detail and commitment to quality have ensured that the book maintains high academic standards.

Our sincere appreciation goes to all the researchers, practitioners, and extension professionals who have shared their experiences and knowledge, helping to bridge the gap between theory and practice. Your contributions have significantly enhanced the practical relevance of this book.

We acknowledge the support of our families and friends who have been a source of strength and inspiration throughout this project. Your patience and understanding have been invaluable.

Lastly, we would like to thank the publishers for their cooperation and professionalism in managing the publication process. Your efforts have ensured that this book reaches a wide audience, contributing to the global discourse on agricultural extension education.

We extend our deepest thanks to everyone involved in this project, directly or indirectly. We hope that this book will serve as a valuable resource for students, researchers, and professionals in the field of agricultural extension, fostering innovation and collaboration in addressing the challenges and opportunities of agricultural development worldwide.

*R. K. Doharey*
*Aman Verma*
*Satyapriya*
*Amrit Warshini*
Editors

# Chapter 1

# Reimagining Extension in a Globalized World

**Aman Verma[1,*]**
**R. K. Doharey[1]**
**Ramesh Chand Bunkar[2]**
**Suresh Kumar Bishnoi[3]**
**Gaurav Kumar[4]**
**Amrit Warshini[1]**
**and Yash Pateriya[1]**

[1]Department of Extension Education, Acharya Narendra Deva University of Agriculture and Technology, Kumarganj, Ayodhya, U.P., India
[2]Division of Dairy Extension, Indian Council of Agricultural Research, National Dairy Research Institute, Haryana, India
[3]Division of Agricultural Extension, Indian Council of Agricultural Research, Indian Agricultural Research Institute, New Delhi, India
[4]Department of Extension Education, Manglayatan University, Jabalpur, Madhya Pradesh, India

## Abstract

Globalization has profoundly impacted agriculture, presenting opportunities and challenges for agricultural extension services. This article explores the need to reimagine extension education in this globalized context. It examines the historical evolution of extension approaches, theoretical frameworks, technological innovations, and the role of global challenges and opportunities. The article proposes new models and methods for agricultural extension education, emphasizing

---

* Corresponding Author's Email: vermaman798577@gmail.com

In: Global Horizons in Agricultural Extension Education
Editors: R. K. Doharey, Aman Verma, Satyapriya et al.
ISBN: 979-8-89530-446-4

context-specific, farmer-centered, technology-enabled, and collaborative strategies. It also highlights the importance of policy and institutional support, including the roles of international organizations and collaborations. Finally, the article discusses future directions, emerging trends, and practical recommendations for policymakers, extension practitioners, and researchers to implement a reimagined extension system that contributes to a more equitable and sustainable global food system.

**Keywords:** agricultural extension, globalization, farmer-centered, technology-enabled, collaboration, policy support, future directions, sustainable agriculture, food security

## 1.1. Introduction

"Agricultural extension facilitates the adoption of improved technologies by assisting farmers in becoming aware of and applying relevant innovations to enhance production efficiency, income, and welfare (World Bank Group). It applies scientific research and agronomic techniques to practice through farmer education, especially for smallholders in developing countries, improving their productivity, food security, and livelihoods (Syngenta Foundation for Sustainable Agriculture)."

**Table 1.1.** The role of agricultural extension in a pre-globalized world

| Role | Description |
|---|---|
| Knowledge dissemination | Agricultural extension services provided farmers with the latest research and knowledge about farming practices. |
| Empowerment | Extension services empowered farmers to make informed decisions about their farming practices by providing them with knowledge and skills. |
| Advancement of agricultural development | Extension services played a crucial role in advancing agricultural development by promoting the adoption of improved farming practices. |
| Risk management | Extension services helped farmers manage risks associated with farming, such as pests, diseases, and weather events. |

In a pre-globalized world, agricultural extension is crucial in disseminating knowledge, empowering farmers, and advancing agricultural development (IFPRI, 2020). The effectiveness of these roles was greatly improved by integrating technology (PLOS ONE, 2023). Agricultural extension was critical to the livelihoods of more than a billion small-scale farmers and other rural people in developing countries. Extension and advisory services provided farmers with advice and information, facilitated innovations and relationships, and assisted in managing risks and disasters (IFPRI, 2020). Table 1.1 summarizes the key roles of agricultural extension in a pre-globalized world, including knowledge dissemination, farmer empowerment, agricultural advancement, and risk management.

## 1.2. Brief Overview of the Current State of Agricultural Extension Education

Nowadays, agricultural extension education is a dynamic and evolving field. It is the application of scientific research and new knowledge to agricultural practices through farmer education (FAO, 2020). The significant activities of agrarian extension education involve technology assessment, demonstration, and capacity development through a network of Agricultural Technology Application Research Institutes (ATARIs) and Krishi Vigyan Kendra (KVKs) (ICAR).

In India, for instance, there are five major agricultural extension systems devoted to extension: the Ministry of Agriculture at the central level, including the Indian Council of Agricultural Research (ICAR) and the Directorate of Extension (DoE); State Departments of Agriculture (DoA), as well as the State Agricultural Universities (SAUs); the Departments of Agriculture (DoA), Animal Husbandry (DAH), Horticulture (DoH) and Fisheries (DoF), as well as the Krishi Vigyan Kendra (KVKs) and, more recently, the Agricultural Technology Management Agency (ATMA) at the District level; also, there are a wide variety of producers groups, including cooperatives and federations of milk, fruits, cotton, oilseeds, coconut, spices etc.; as well as civil (ICAR).

## 1.3. The Need for Reimagining Extension in a Globalized World

The world is increasingly interconnected, with exports as a proportion of GDP having doubled since 1990, intangibles like data and IP having grown twice as fast as goods since 2010, and cross-border talent migration at an all-time high (McKinsey & Company, 2022). This hyper-connectivity has put globalization under question more than it ever has been. Between the massive disruptions caused by the pandemic and the war in Ukraine, the world is under pressure to deglobalize and instead move towards localization, not being interdependent and sharing friends, to name a few concepts that have come up in this discourse.

However, the reality of our world is interconnected, and globalization has, by and large, been a force for good, increasing education levels and GDP growth and reducing poverty, particularly in low- and middle-income countries (McKinsey & Company, 2022). Therefore, there is a need to reimagine agricultural extension education in this globalized world. Transporting physical goods across borders is no longer the only, or even the primary, driving force behind global integration. Instead, we are increasingly connected by flows of intangibles, services, and talent (McKinsey & Company, 2022).

As such, agricultural extension education must evolve to meet these new challenges and opportunities. It needs to leverage technology and innovation to improve farming practices and knowledge among farmers, mainly smallholders in developing countries. It must also foster cross-border cooperation and knowledge sharing to address global challenges such as food security and climate change.

## 1.4. Highlighting the Rise of Globalization and Its Impact on Agriculture

Globalization, characterized by an interchange of worldviews, products, ideas, and cultural aspects, has significantly impacted agriculture. It has increased the movement of goods, services, capital, technology, information, and people across borders (Joginder, 2023). This has resulted in the introducing of new seeds, fertilizers, and technologies in the agricultural sector, potentially replacing traditional methods and inputs. However, these changes have also brought challenges, such as reduced soil fertility, increased water usage, and unemployment (Mohta, 2020).

Globalization has further influenced crop choices. Farmers have shifted towards crops with higher international market demand. While economically beneficial, this shift may lead to the neglect of native crops and biodiversity (Joginder, 2023).

## 1.5. Introducing the Concept of "Reimagining Extension" in This New Context

Agricultural extension, in essence, refers to the formal education system designed to improve farming practices and knowledge among farmers (FAO, 2014). In a globalized agricultural landscape, "reimagining extension" becomes crucial.

Reimagining extension involves adapting and innovating extension strategies to meet the challenges and opportunities of globalization. It consists of facilitating farmers' access to knowledge, information, and technologies, facilitating their interaction with partners in research, education, agribusiness, and other relevant institutions, and assisting them in developing their own technical, organizational, and management skills and practices.

In this new context, extension services need to be boosted through timely recruitment, periodic training of agents, and provision of adequate logistics. The goal is to enhance farm productivity and household income, contributing to sustainable agricultural development in a globalized world.

## 1.6. Historical Context and Evolution

### 1.6.1. Brief History of Agricultural Extension Services

Agricultural extension, the formal education system designed to improve farming practices and knowledge among farmers, has a long history (FAO, 2014). "extension" originated in England in 1866. However, the birth of the modern extension service is attributed to a crisis in Ireland during the mid-19th Century. A devastating potato famine (1845-1851) caused by fungal disease led to the establishment of the first modern agricultural extension service.

### 1.6.2. Shifts Influenced by Major Global Events

The Green Revolution, a period of agricultural transformation focused on increased productivity through modern techniques, was crucial in addressing hunger and poverty in many countries during the mid-20th Century (Britannica, 2009; Spanne, n.d.; Desk, 2023; Sadaf, 2023). Currently, a similar large-scale shift, the "global green shift," emphasizes technological and economic changes (Fagerberg, 2023).

### 1.6.3. Transition from Local to Global Extension Approaches

Extension service delivery has evolved from "push" approaches, where research findings were disseminated by government agents, to "pull" strategies that focus on farmer needs, and finally to "innovation systems" approaches where intermediary groups connect farmers, markets, researchers, and extension workers (FAO, 2018). These transformations, summarized in Table 1.2, highlight the progressive shifts in extension services over time. As a response to global efforts in food systems, the African Forum for Agricultural Advisory Services (AFAAS) has developed the Agroecological Transition, Responsive Extension Approaches (ATREA) program (AFAAS, n.d.).

**Table 1.2.** Overview of the evolution of agricultural extension services over time

| Period | Extension Approach | Key Characteristics |
|---|---|---|
| Early 20$^{th}$ Century | Push Based | Research shared through government extension officers |
| Mid 20$^{th}$ Century | Pull Based | Extension based on farmer needs |
| Late 20$^{th}$ Century to Present | Innovation Systems | Groups formed as intermediaries between farmers, market suppliers, buyers, extension workers, and researchers. |

## 1.7. Theoretical Frameworks

Theoretical frameworks provide a basis for understanding and interpreting phenomena systematically. Several theories have been instrumental in shaping contemporary practices in the context of extension practices.

### 1.7.1. Diffusion of Innovations Theory

The Diffusion of Innovations theory, proposed by Everett Rogers, provides a model for understanding how ideas and technology spread through societies

(Rogers, 2003). This theory has been applied in extension practices to facilitate the adoption of new agricultural technologies and practices among farmers. However, in a globalized context, the effectiveness of this theory can be limited by cultural, economic, and institutional barriers.

### 1.7.2. Adult Learning Theory

Adult Learning Theory, or Andragogy, emphasizes the value of the learning process. It uses an approach to learning that is problem-based and collaborative rather than didactic, and it also emphasizes equality between the teacher and learner (Knowles, 1984). This theory is particularly relevant in extension practices where farmers, as adult learners, are encouraged to take an active role in their learning process. In a globalized context, the effectiveness of this theory can be influenced by the learners' cultural background and previous educational experiences.

### 1.7.3. Systems Theory

Systems Theory views organizations as systems of interconnected parts that function together to achieve a common goal (Bertalanffy, 1968). This theory is used in extension practices to understand the complex interactions between different stakeholders in agricultural systems. In a globalized context, the effectiveness of this theory can be challenged by the complexity and diversity of global agricultural systems.

### 1.7.4. Application of Theories in Contemporary Extension Practices and Their Effectiveness in a Globalized Context

Applying the theories above—Diffusion of Innovations, Adult Learning, and Systems Theory—to contemporary extension practices has significantly shaped how knowledge is disseminated and adopted. However, their effectiveness in a globalized context warrants critical analysis.

### 1.7.5. Application of Theories

The *Diffusion of Innovations theory* has been instrumental in understanding how new agricultural practices or technologies are adopted within a community (Rogers, 2003). Extension services have used this theory to design interventions that facilitate adoption, such as demonstration plots or farmer field schools.

The *Adult Learning theory* has guided the development of extension programs that are learner-centered and problem-oriented, recognizing that adult learners have unique needs and preferences (Knowles, 1984). This has led to using participatory approaches in extension, where farmers are actively involved in learning.

*Systems Theory* has been applied in extension to understand the complex interactions between various components of agricultural systems (Bertalanffy, 1968). This has informed the design of holistic and integrated extension programs that consider the interdependencies of different system elements.

### 1.7.6. Critical Analysis in a Globalized Context

In a globalized context, the effectiveness of these theories is influenced by various factors. For instance, the Diffusion of Innovations theory may not account for cultural differences in the perception and adoption of innovations. Similarly, differing educational backgrounds and learning preferences across cultures may challenge the Adult Learning theory. The Systems Theory, while providing a holistic view, may struggle with the complexity and diversity of global agricultural systems.

Therefore, while these theories provide valuable insights, their application in a globalized context should be based on understanding the local context and adaptability to each situation's unique characteristics.

## 1.8. Technological Innovations in Extension

Technological innovations have significantly influenced extension services, reshaping how information is disseminated and how agricultural practices are conducted.

### 1.8.1. Role of ICT in Extension

Information and Communication Technology (ICT) is pivotal in extension services. It facilitates the dissemination of information, enhances communication between farmers and extension agents, and improves data collection and analysis (Aker, 2011).

### 1.8.2. Disruptive Technologies in Agriculture

Several disruptive technologies are reshaping agricultural practices:

- *Artificial Intelligence (AI):* AI can analyze large amounts of data to provide insights on crop health, pest infestation, and yield prediction, thereby aiding decision-making processes (Rotolo et al., 2018).
- *Remote Sensing:* Remote sensing technologies, such as drones and satellites, provide real-time data on weather conditions, soil health, and crop growth. This information can be used to optimize farming practices (Thenkabail, 2019).
- *Blockchain*: Blockchain technology can enhance traceability in the agricultural supply chain, ensuring food safety and reducing fraud (Kamble et al., 2019).

### 1.8.3. Barriers to Technology Adoption

Despite the potential benefits, several barriers hinder the adoption of these technologies:

- *Infrastructure:* Lack of necessary infrastructure, such as internet connectivity and electricity, can limit the use of these technologies (Wyche et al., 2016).
- *Digital Literacy:* Farmers' lack of digital literacy can also hinder technology adoption (Medlin et al., 2018).
- *Cost:* The high cost of some technologies can be prohibitive for small-scale farmers (Chavula, 2014).

### 1.8.4. Strategies to Overcome Barriers

Several strategies can be employed to overcome these barriers:

- *Capacity Building:* Training programs can be conducted to improve farmers' digital literacy (Medlin et al., 2018).
- *Public-Private Partnerships*: Such partnerships can help develop the necessary infrastructure (Chavula, 2014).
- *Subsidies and Loans:* Providing subsidies or low-interest loans can make technologies more affordable for farmers (Chavula, 2014).

## 1.9. Global Challenges and Opportunities: A Detailed Examination of Current Global Challenges

### 1.9.1. Climate Change

Climate change poses a significant threat to agriculture, impacting all aspects of food security: availability, access, utilization, and stability (IPCC, 2019). Rising temperatures and changes in rainfall patterns can negatively affect yields of crops, livestock, and fish (IPCC, 2019). Additionally, climate change can increase the prevalence of pests and diseases, disrupt markets, and reduce the purchasing power of low-income consumers for food (World Bank, 2022).

### 1.9.2. Food Security

Food security is another pressing global issue. The number of people facing acute food insecurity has risen dramatically in recent years due to factors like conflict, supply chain disruptions, and the economic impacts of the COVID-19 pandemic (World Bank, 2022). Rising food commodity prices have pushed millions more people in low-income countries towards food insecurity (World Bank, 2022).

### 1.9.3. Sustainability

Balancing the growing demand for food with environmental resource conservation is a complex challenge in sustainable agriculture. In the 21st Century, global food systems face the dual challenge of increasing food production while competing for limited resources like land, water, and energy, significantly impacting the food supply (Columbia University, n.d.). As Figure 1.1 illustrates the interconnectedness of these three critical global challenges.

**Figure 1.1.** Interconnectedness of three critical global challenges: Climate Change, Food Security, and Sustainability.

Here's a breakdown of the arrows and their implications:

- *Climate Change to Food Security*: This arrow suggests climate change directly impacts food security. Climate changes can affect food production, availability, and distribution, leading to challenges in global food demand.
- *Climate Change to Sustainability*: This arrow indicates climate change influences sustainability efforts. It highlights the need for sustainable practices to mitigate the effects of climate change on the environment and human societies.

- *Food Security to Sustainability*: The arrow from Food Security to Sustainability implies that ensuring a stable and sufficient food supply is essential for sustainable development. It emphasizes the importance of sustainable agricultural practices to maintain food security.
- *Outward Arrows to Opportunities*: The arrows pointing outwards from each circle to the central 'Opportunities' circle represent the potential for technological advancements, improved resource management, and innovative approaches to address these challenges and turn them into opportunities for progress.

## 1.10. Opportunities Presented by Globalization

Globalization has been a driving force in shaping modern economies and societies. It has facilitated the spread of technology, fostered multinational partnerships, and introduced innovative funding mechanisms. Each of these aspects plays a crucial role in the interconnected world economy.

### 1.10.1. Technology Transfer

The transfer of technology across borders has intensified due to globalization. This movement has been particularly beneficial in emerging markets, where it has helped to boost innovation and productivity. For instance, using artificial intelligence in healthcare in China exemplifies how technology can spread rapidly and improve living standards (IMF, 2018). Globalization accelerates development and enables countries to leapfrog to more advanced technologies without developing them indigenously.

### 1.10.2. Multinational Partnerships

Globalization has paved the way for multinational partnerships, which are collaborations between companies, governments, and other entities across different countries. These partnerships can lead to pooled resources, shared expertise, and access to new markets. For example, global value chains have become a vehicle for technology transfer, allowing firms in developing countries to assess and implement foreign technology, thereby enhancing their performance (Rigo, 2020).

### 1.10.3. Innovative Funding Mechanisms

Innovative funding mechanisms have emerged as a response to the need for financing development beyond traditional methods. These mechanisms include solidarity levies, such as small taxes added to airline tickets, and the International Finance Facility for Immunization, which uses the borrowing capacity of states to finance immunization programs in various countries (United Nations, 2023). As Table 1.3 summarizes, globalization presents significant opportunities to address key challenges such as climate change, food security, and sustainability.

**Table 1.3.** Summary of challenges and opportunities

| Challenge | Description | Opportunity Presented by Globalization |
|---|---|---|
| Climate Change | Rising temperatures and changing rain patterns threaten agriculture. | Technology transfer for climate-smart practices. |
| Food Security | The growing population strains food production systems. | Innovation in agricultural technology and resource management. |
| Sustainability | Balancing food production with environmental resources. | Multinational partnerships to promote sustainable practices. |

## 1.11. Case Studies Showcasing Successful Integration of Global Opportunities into Extension Services

Agricultural extension services play a pivotal role in modernizing and sustaining agriculture. They act as a bridge between research and farming practices, disseminating knowledge and innovations to increase productivity and improve livelihoods. The integration of global opportunities into extension services has been a critical factor in the success of agricultural development in various regions. Here, we explore a few case studies that highlight this integration.

### 1.11.1. Demand-Driven Approaches in Benin

In Benin, a shift towards community-driven development has led to faster and more cost-effective outcomes. Integrating local knowledge with global agricultural practices has empowered communities to take charge of their development, leading to significant improvements in agricultural productivity (World Bank, 2003).

### 1.11.2. Farmer Research Committees in Colombia

Colombia has seen the spread of Local Agricultural Research Committees (CIALS), which emphasize extension through farmer research. This approach has successfully integrated farmers into the research process, ensuring that the solutions developed are well-suited to local conditions and have a higher adoption rate (Quirós Torres et al., 2003).

### 1.11.3. Farmer Field Schools in West Africa

The Farmer Field School (FFS) approach has been an effective extension strategy in West Africa. It has provided farmers with hands-on experience and learning opportunities, leading to better crop management and increased yields. The FFS approach has successfully incorporated global best practices in sustainable agriculture into local farming systems (Owens & Simpson, 2003).

### 1.11.4. Digital Technologies in Agricultural Extension

The rapid proliferation of digital technologies offers exciting opportunities to bridge the gap between extension services and farmers. Agricultural extension can leverage mobile applications, SMS-based advisory services, and interactive voice response systems to disseminate information directly to farmers' smartphones, thus integrating global technological advancements into local agricultural practices (Just Agriculture, 2023).

## 1.12. Reimagining Extension Education

### 1.12.1. Proposing New Models and Approaches for Agricultural Extension Education in a Globalized World

Agricultural extension, in its essence, refers to the formal education system designed to improve farming practices and knowledge among farmers (FAO, 2014). However, the traditional top-down, supply- and technology-driven extension approach is no longer deemed appropriate in the face of the complex challenges of a globalized world (Danso-Abbeam et al., 2024).

New models and approaches are being proposed and implemented in response to these challenges. For instance, the Sub-Saharan Africa region has seen the development of various extension models, including the National Public Extension Model, Training and Visit (T&V) Extension Model, Private Sector Model, Fee-For-Service Extension Models, Farmer Field School Model, Non-Governmental Organization Extension Model, and Commodity Extension and Research Model (Sennuga et al., 2024). These models are structured around key themes such as top-down, participatory, demand-led, group versus individual targeting, private sector, and free/paid extension services (Sennuga et al., 2024).

### 1.12.2. Discussing the Role of Stakeholders in this Reimagination

Stakeholders play a significant role in the overall development of an institution or organization. Their contribution and collaboration make the program effective, acting as motivators, decision-makers, and sources of inspiration and hope (Stakeholders in Education, n.d.). In agricultural extension education, stakeholders include governments, educational institutions, communities, and more.

Governments and policymakers are responsible for developing and implementing educational policies that shape the entire system. They work alongside educators, administrators, and stakeholders to create a comprehensive framework that addresses the diverse needs of learners (Davis, 2020)

Educational institutions, particularly universities and research institutes, play a pivotal role in generating new knowledge and technologies, which are then disseminated to farmers through extension services (FAO, 2014).

Communities, especially farming communities, are both the beneficiaries and contributors to agricultural extension education. They provide valuable insights into local needs and challenges, which can guide the development of context-specific extension services (FAO, 2014).

In conclusion, reimagining agricultural extension education in a globalized world involves proposing new models and approaches and engaging various stakeholders. This collective effort can lead to more effective and sustainable farming practices, ultimately contributing to global food security and rural development.

## 1.13. Reimagining Extension for the 21st Century

The 21st Century presents a unique opportunity to reimagine agricultural extension services. This reimagining involves a shift towards more context-specific and farmer-centered approaches. As shown in Table 1.4, these approaches emphasize tailored services, farmer participation, digital tools, and multi-stakeholder collaboration, each contributing to increased relevance, innovation, and knowledge dissemination.

### 1.13.1. Context-Specific: Tailoring Extension Services to Local Needs and Cultures

Context-specific extension services involve tailoring these services to local needs and cultures. This approach recognizes that agricultural practices and challenges can vary significantly across regions and cultures. Consequently, extension services necessitate flexibility and adaptability to effectively cater to these distinct contexts (Guildhawk, 2023). For example, extension professionals might offer tailored advice considering local obstacles and requirements within a particular field or industry (Danso-Abbeam et al., 2024). This approach ensures that the extension services are relevant and beneficial to the local farmers, improving their productivity and profitability.

### 1.13.2. Farmer-Centered: Empowering Farmers as Active Participants in Knowledge Creation and Sharing

Farmer-centered extension services empower farmers to participate actively in knowledge creation and sharing. This approach recognizes farmers as valuable sources of knowledge and innovation rather than merely recipients of information (Ahmed & Kirtuniya, 2024).

For example, farmer-to-farmer knowledge-sharing initiatives can catalyze enhancing farming practices and bolstering community resilience. Moreover, Farmer-Centered Multi-stakeholder Platforms can create multifunctional spaces for peer-to-peer learning, interaction with institutional actors, and network building, empowering farmers to become architects of their prosperity (Van Ewijk et al., 2024).

### 1.13.3. Technology-Enabled: Utilizing Digital Tools for Information Dissemination and Communication

The advent of digital technology has revolutionized the field of agricultural extension. Digital communication tools such as mobile health apps, telemedicine, and online health information resources have demonstrated significant potential to improve health literacy, ultimately leading to better health outcomes (Fitzpatrick PJ, 2023). These tools facilitate the access of farmers, their organizations, and other market actors to knowledge, information, and technologies. They also encourage interaction with partners in research, education, agribusiness, and other relevant institutions (Danso-Abbeam et al., 2024).

For instance, the rise in digital information dissemination has led to new trends in communication mediums, with a particular push towards new avenues that promote quicker dissemination of information earlier in the approval process (AMCP.org, 2021). This has allowed for the effective management of natural resources and coordination of farming practices (Enago Academy, 2024).

### 1.13.4. Collaborative: Building Partnerships between Researchers, Extension Agents, Farmers, and Other Stakeholders

Collaboration between various stakeholders in the agricultural sector is crucial for sustainable agriculture. More and better collaboration between farmers and other stakeholders has repeatedly been identified as a key strategy for sustainable agriculture (Velton, 2021). This involves building partnerships between researchers, extension agents, farmers, and other stakeholders (Dosso 2024).

For instance, farmer field schools provide a fruitful learning environment where researchers, extension workers, and farmers can collaboratively test innovative climate change adaptation practices and technologies (FAO). Such collaborative efforts can generate ecological, social, and economic benefits in agricultural contexts (Velton, 2021).

**Table 1.4.** Reimagined extension - Key components and benefits

| Component | Description | Benefits |
|---|---|---|
| Context-Specific Services | Tailored to local needs and cultures | Increased relevance, improved farmer adoption |
| Farmer-Centered Activities | Empower farmers in knowledge-sharing | Increased innovation, improved farmer ownership |
| Technology-Enabled Tools | Facilitate information access and communication | Improved knowledge dissemination, wider reach |
| Collaborative Partnerships | Researchers, extension agents, and farmers work together | Enhanced problem-solving, co-creation of solutions |

## 1.14. Policy and Institutional Support

Policy frameworks play a crucial role in shaping the effectiveness of extension services in a globalized environment. Depending on their design and focus, they can either facilitate or hinder the implementation of extension programs.

### 1.14.1. Policy Frameworks Supporting Extension Services

Policies supporting extension services often promote innovation, facilitate information access, and foster stakeholder collaboration. For instance, policies that encourage using Information and Communication Technology (ICT) in extension services can enhance their reach and efficiency (Aker, 2011). Similarly, policies that promote collaboration among farmers, researchers, and extension agents can facilitate the exchange of knowledge and best practices (Sulaiman & Davis, 2012).

## 1.14.2. Policy Frameworks Hindering Extension Services

Conversely, certain policy frameworks can hinder the effectiveness of extension services. For example, policies that restrict the use of certain technologies can limit the potential of extension services to leverage these technologies for information dissemination and data collection (Chavula, 2014). Additionally, policies that fail to address smallholder farmers' unique needs and circumstances can limit their ability to benefit from extension services (Anderson & Feder, 2004).

## 1.14.3. Role of International Organizations and Collaborations

International organizations and collaborations play a pivotal role in shaping and supporting extension services across the globe. They provide a platform for knowledge exchange, capacity building, and policy advocacy, thereby enhancing the effectiveness of extension services.

### *1.14.3.1. International Organizations*

Organizations such as the Food and Agriculture Organization (FAO), the World Bank, and the International Fund for Agricultural Development (IFAD) have supported extension services. They provide technical assistance, funding, and policy guidance to countries, helping them to develop and implement effective extension programs (Sulaiman & Davis, 2012).

### *1.14.3.2. Collaborations*

Regional and global collaborations facilitate the sharing of knowledge, best practices, and resources among countries. They enable countries to learn from each other's experiences and adapt successful strategies to their contexts. Collaborations foster innovation by bringing diverse perspectives and expertise (Rivera & Sulaiman, 2009).

#### *1.14.3.3. Recommendations for Policy Makers*

To enhance the scope and effectiveness of extension services, policymakers could consider the following recommendations:

- *Promote ICT Use:* Policies should promote Information and Communication Technology (ICT) in extension services, as it can facilitate information dissemination and enhance communication (Aker, 2011).
- *Invest in Infrastructure*: Investments should be made in infrastructure, such as internet connectivity and electricity, to facilitate the use of ICT in extension services (Chavula, 2014).
- *Enhance Digital Literacy*: Programs should be implemented to enhance farmers' digital literacy, as this can facilitate the adoption of ICT in extension services (Medlin et al., 2018).
- *Encourage Collaboration*: Policies should encourage collaboration among farmers, researchers, and extension agents to facilitate the exchange of knowledge and best practices (Sulaiman & Davis, 2012).
- *Support Research and Innovation*: Policies should support research and innovation in agriculture to develop new technologies and practices that can enhance the effectiveness of extension services (Feder et al., 2001). This table is a simplified representation; each policy recommendation's impact can vary significantly depending on each extension service's specific context and characteristics.

## 1.15. Future Directions

The future of agricultural extension is likely to be shaped by several emerging trends and global economic factors. Here are some insights into these aspects:

### 1.15.1. Merging Trends in Agricultural Extension

- *Participatory Approaches:* There is a growing trend towards more participatory approaches in extension services. These approaches involve farmers and local communities in decision-making, ensuring services are tailored to their needs and circumstances (Chambers, 1994).

- *Focus on Youth and Women:* An increasing focus is on engaging youth and women in agricultural extension. This is driven by recognizing crucial role in agriculture and the need to empower them with the necessary knowledge and skills (FAO, 2018).

### 1.15.2. Potential Impacts of Global Economic Trends

Global economic trends can significantly impact extension services. For instance, the increasing globalization of agricultural markets can create opportunities and challenges for extension services. On the one hand, it can open new markets for farmers, but on the other hand, it can also expose them to greater competition and market volatility (World Bank, 2007).

### 1.15.3. Predictions for the Future Role of Extension Agents and Educators

The role of extension agents and educators will likely evolve in response to these trends and challenges. They will need to play a more facilitative role, helping farmers to navigate the complexities of global markets and to adopt sustainable farming practices. They must also embrace new technologies and participatory approaches to deliver more effective and inclusive services (Sulaiman & Davis, 2012).

## 1.16. Recommendations

To implement a reimagined extension system, policymakers, extension practitioners, and researchers could consider the following practical recommendations:

### 1.16.1. For Policymakers

- *Promote ICT Use:* Policies should encourage using Information and Communication Technology (ICT) in extension services, as it can facilitate information dissemination and enhance communication.

- *Invest in Infrastructure:* Policymakers should prioritize investments in infrastructure, such as internet connectivity and electricity, to facilitate the use of ICT in extension services.
- *Support Research and Innovation:* Policies should support research and innovation in agriculture to develop new technologies and practices that can enhance the effectiveness of extension services.

### 1.16.2. For Extension Practitioners

- *Embrace Participatory Approaches*: Extension practitioners should embrace more participatory approaches involving farmers and local communities in decision-making.
- *Focus on Youth and Women:* Extension services should focus on engaging youth and women, recognizing their crucial role in agriculture and the need to empower them with the necessary knowledge and skills.
- *Leverage New Technologies*: Extension practitioners should use new technologies, such as AI, remote sensing, and blockchain, to enhance their services.

### 1.16.3. For Researchers

- *Conduct Context-Specific Research*: Researchers should conduct context-specific research to understand different farming communities' unique needs and circumstances.
- *Collaborate with Practitioners:* Researchers should collaborate closely with practitioners to ensure their research is relevant and applicable.
- *Evaluate Impact:* Researchers should evaluate the impact of different extension approaches and technologies to identify best practices and areas for improvement.

## Conclusion

In conclusion, the evolving realities of globalization and the adaptable approach to agricultural extension, integrate theoretical frameworks, technological innovations, supportive policy and institutional structures form a cohesive vision for the future of extension services. The emerging trends of participatory approaches and a focus on youth and women, coupled with the potential impact of global economic trends, are reshaping the role of extension agents and educators. The reimagined extension system, underpinned by continual adaptation and innovation, holds immense potential to contribute to a more equitable and sustainable global food system. Therefore, it is imperative for all stakeholders in agricultural extension - policymakers, practitioners, and researchers - to embrace change and innovation and work collaboratively towards realizing this vision.

## References

Abhishek, A. (2023). *Describe The Impact of Globalisation on Indian Agriculture.* Retrieved April 14, 2024, from https://agriculturereview.com/impact-of-globalisation-on-indian-agriculture/.

aesanetwork.org.

African Forum for Agricultural Advisory Services (AFAAS). (n.d.). *Agroecological Transition, Responsive Extension Approaches (ATREA).* https://www.afaas-africa.org/category/atrea/.

*Agricultural extension (English).* IEG lessons and practices Washington, D.C.: World Bank Group. http://documents.worldbank.org/curated/en/956441468763806805/Agricultural-extension.

agriwelfare.gov.in.

Ahmed, A., & Kirtuniya, S. (2024). Empowering Farming Communities through Farmer-to-Farmer Knowledge Sharing.

Aker, J. C. (2011). Dial "A" for agriculture: a review of information and communication technologies for agricultural extension in developing countries. *Agricultural Economics*, 42(6), 631-647.

AMCP.org. (2021). The Rise in Digital Information Dissemination: Communication Trends in Managed Care.

Anderson, J. R., & Feder, G. (2004). Agricultural extension: Good intentions and hard realities. *The World Bank Research Observer*, 19(1), 41-60.

Bertalanffy, L. V. (1968). *General System Theory: Foundations, Development, Applications*. New York: George Braziller.

Britannica, The Editors of Encyclopaedia. (2009, March 12). Green revolution | Definition, advantages, importance, & facts | Britannica. https://www.britannica.com/facts/green-revolution.

Chambers, R. (1994). *Participatory rural appraisal (PRA): Analysis of experience*. World Development, 22(9), 1253-1268.

Chavula, H. K. (2014). The role of ICTs in agricultural production in Africa. *Journal of Development and Agricultural Economics*, 6(7), 279-289.

Columbia University. (n.d.). Future of Food: Exploring Challenges to Global Food Systems – State of the Planet. https://www.climate.columbia.edu/events/state-worlds-food-system-webinar.

Danso-Abbeam, G., Ehiakpor, D. S., & Aidoo, R. (2018). Agricultural extension and its effects on farm productivity and income: insight from Northern Ghana[7].

Danso-Abbeam, G., Ehiakpor, D. S., & Aidoo, R. (2024). Advisory Services: Extension workers serve as advisors, offering personalised guidance based on people's specific needs and challenges as per the needs of the field/sector. IJCRT.

Davis, Kristin E., ed.; Babu, Suresh Chandra, ed.; and Ragasa, Catherine, ed. 2020. Agricultural extension: Global status and performance in selected countries. Washington, DC: International Food Policy Research Institute (IFPRI). https://doi.org/10.2499/9780896293755.

Desk, E. (2023). The Green Revolution: Agricultural Transformation in Asia and Latin America - Scale Climate Action. Scale Climate Action. https://ianas.org/wp-content/uploads/2022/01/Climate-Variabillity-in-Cnetral-America-What-does-it-mean-for-Water-Managers.pdf.

Dosso, Faridah, Gouroubera, W, M., Idrissou, Latifou, Moumouni-mousa, & Ismail. (2024). The combination of extension approaches strengthens farmers' innovativeness in sustainable land managemet. *Environment, Development and Sustainability, 26*(4), 10043-10062.

Enago Academy. (2024). Effective Use of Tables and Figures in Research Papers.

*Enhancing linkages between extension, research and producers through innovations.* https://www.fao.org/3/cb2110en/CB2110EN.pdf.

Fagerberg, J. (2023). *The global green shift.* https://www.janfagerberg.org/the-global-green-shift/.

FAO. (2014). Agricultural extension: Global status and performance in selected countries. https://www.ifpri.org/interactive/ag-extension.

FAO. (2014). Agricultural extension: Global status and performance in selected countries[3].

FAO. (2018). *Global review of extension approaches and models.* https://www.semanticscholar.org/paper/GLOBAL-REVIEW-OF-EXTENSION-APPROACHES-AND-MODELS/c705038b76dbe341a8e2859688412f17e7e146de.

FAO. (2018). *Youth and agriculture.* Rome: Food and Agriculture Organization of the United Nations.

FAO. (2020). Agricultural and Rural Extension Worldwide. ICAR. Agricultural Extension Division.

Feder, G., Murgai, R., & Quizon, J. B. (2001). The acquisition and diffusion of knowledge: The case of pest management training in farmer field schools, Indonesia. *Journal of Agricultural Economics*, 53(2), 221-243.

Fitzpatrick PJ (2023) Improving health literacy using the power of digital communications to achieve better health outcomes for patients and practitioners. Front. Digit. Health 5:1264780. doi: 10.3389/fdgth.2023.1264780.

Guildhawk. (2023). Tailoring content to reach global markets: The power of localization.

Guildhawk. (2023). Tailoring content to reach global markets: the power of localization. *Guildhawk.* https://www.guildhawk.com/blog/tailoring-content-to-reach-global-markets-the-power-of-localization.

icar.gov.in.

IFPRI. (2020). Agricultural extension: Global status and performance in selected countries[4]

IMF. (2018). Globalization Helps Spread Knowledge and Technology Across Borders. https://www.imf.org/en/Blogs/Articles/2018/04/09/globalization-helps-spread-knowledge-and-technology-across-borders.

Intergovernmental Panel on Climate Change (IPCC). (2019). *Climate Change and Land: An IPCC Special Report on climate change, desertification, land degradation, sustainable land management, food security, and greenhouse gas fluxes in terrestrial ecosystems.* https://www.ipcc.ch/srccl/chapter/chapter-5/.

Joginder, S. (2023). The impact of globalization on agriculture: A review. *Journal of Agriculture and Sustainability, 8*(2), 1-15.

Just Agriculture. (2023). Bridging the Gap: Integrating Digital Technologies in Agricultural Extension.

Kamble, S. S., Gunasekaran, A., & Arha, H. (2019). Understanding the Blockchain technology adoption in supply chains-Indian context. *International Journal of Production Research*, 57(7), 2009-2033.

Knowles, M. S. (1984). *Andragogy in Action*. San Francisco: Jossey-Bass.

krishi.icar.gov.

McKinsey & Company. (2022). Reimagining Globalization and Growth.

Medlin, B. D., Cazier, J. A., & Foulk, D. (2018). Digital literacy and the digital divide: Information behavior and seeking of low socioeconomic students. *Information Systems Education Journal*, 16(3), 6.

Mohta, R. (2020). Globalization and its effects on Indian agriculture sector[2].

Mohta, S. (2020). Globalization and agricultural development: A complex relationship. *International Journal of Agricultural Economics and Policy*, 5(2), 12-21.

Owens, M., & Simpson, B. M. (2003). *West Africa: Farmer Field Schools as an Extension Strategy.*

*PLOS ONE.* (2023). A scoping review on technology applications in agricultural extension[5]

Quirós Torres, C. A., Douthwaite, B., Roa Velasco, J. I., & Ashby, J. (2003). Colombia, Latin America and the Spread of Local Agricultural Research Committees (CIALS).

Rigo, D. (2020). Global value chains and technology transfer: new evidence from developing countries. *Review of World Economics*, 157, 271-294.

Rivera, W. M., & Sulaiman, V. R. (2009). Extension: object of reform, engine for innovation. *Outlook on Agriculture*, 38(3), 267-273.

Rogers, E. M. (2003). *Diffusion of Innovations, 5th Edition*. New York: Free Press.

Rotolo, D., Hicks, D., & Martin, B. R. (2018). What is an emerging technology? *Research Policy*, 47(10), 1823-1843.

Sadaf. (2023). Green Revolution - Scheme, Features and Objectives - CareerByWell. CareerByWell. https://www.careerbywell.com/.

Sennuga Samson Olayemi, Alo Adeola Ope-Oluwa, and Angba Cynthia Whiteley. (2024). Evolution of Agricultural Extension Models in Sub-Saharan Africa: A Critical Review. I*nternational Journal of Agricultural Extension and Rural Development Studies* Vol.8, No.1, pp.29-51, 2021.

Spanne, A. (n.d.). *Green Revolution: History, Technologies, and Impact.* Treehugger. https://www.thoughtco.com/green-revolution-overview-1434948.

Sulaiman, V. R., & Davis, K. (2012). The "New Extensionist": Roles, Strategies, and Capacities to Strengthen Extension and Advisory Services. *Agriculture and Rural Development Discussion Paper*, 44.

Syngenta Foundation for Sustainable Agriculture. Agricultural Extension[3]

Thenkabail, P. S. (Ed.). (2019). *Remote Sensing Handbook - Three Volume Set: Land Resources Monitoring, Modeling, and Mapping with Remote Sensing*. CRC Press.

United Nations. (2023). Innovative Financing Mechanisms and Solutions.

van Ewijk, E., Ataa-Asantewaa, M., Asubonteng, K. O., Van Leynseele, Y. P. B., Derkyi, M., Laven, A., & Ros-Tonen, M. A. F. (2024). Farmer-Centered Multi-stakeholder Platforms: From Iterative Approach to Conceptual Embedding.

Velten, S., Jager, N. W. & Newig, J. Success of collaboration for sustainable agriculture: a case study meta-analysis. *Environ Dev Sustain* 23, 14619–14641 (2021). https://doi.org/10.1007/s10668-021-01261-y.

Wikipedia Contributors. (2005, February 04). Agricultural extension - Wikipedia. Wikimedia Foundation. Inc. https://en.wikipedia.org/wiki/Agricultural_extension.

World Bank. (2003). Demand-Driven Approaches to Agriculture Extension. Case Studies of International Initiatives.

World Bank. (2007). *World Development Report 2008: Agriculture for Development.* Washington, DC: World Bank. doi:10.1596/978-0-8213-6807-7. License: Creative Commons Attribution CC BY 3.0 IGO.

World Bank. (2022, October 19). Climate Explainer: Food Security and Climate Change. https://www.worldbank.org/en/news/feature/2022/10/17/what-you-need-to-know-about-food-security-and-climate-change.

Wyche, S., Steinfield, C., Caine, K., Zanella, A., & Simiyu, N. (2016, May). Reflecting on video: Exploring the efficacy of video for teaching device literacy in rural Kenya. In *Proceedings of the Eighth International Conference on Information and Communication Technologies and Development* (pp. 1-11).

Chapter 2

# Agriculture and the SDGs: A Review of Quantitative Methods, Models and Approaches

**Alaka S. Balan[1,*]**
**Sreelekshmy Sadanand[2]**
**V. Sreeram [3]**
**and Navitha Raj[4]**
[1]Department of Agricultural Extension, College of Agriculture, Ambalavayal, Kerala Agricultural University, Kerala, India
[2]Department of Extension Education, College of Agriculture, Punjab Agricultural University, Ludhiana, Punjab, India
[3]Department of Agricultural Extension, Regional Agricultural Research Station, Ambalavayal, Kerala Agricultural University, Kerala, India
[4]Department of Agricultural Extension, College of Agriculture, Ambalavayal, Kerala Agricultural University, Kerala, India

## Abstract

Environmental impacts on agricultural systems and food supplies are significant. For instance, one-third of greenhouse gas (GHG) emissions, up to 80 percent of atrophying emissions, and 86 percent of species at risk of extinction are caused by the agri-food sector. Despite commitments, progress towards achieving the 169 targets of the 17 Sustainable Development Goals (SDGs) stands at a mere 15 percent, with some areas even showing regression. In India, advancements in the agricultural sector continue to expand to meet the food demand of the

[*] Corresponding Author's Email: alakasudhee@gmail.com

In: Global Horizons in Agricultural Extension Education
Editors: R. K. Doharey, Aman Verma, Satyapriya et al.
ISBN: 979-8-89530-446-4

country's growing population. However, due to the massive use of agrochemicals and emission-intensive farming, modern agriculture not only addresses hunger but also has an impact on the ecosystem, human health, and resources, necessitating a sustainable assessment. Furthermore, crop production in India generates a significant amount of biomass waste, a considerable portion of which is either burned or incorporated into the soil through tillage. These practices undermine the economy, the environment, and public health—key pillars of sustainable development—both directly and indirectly. SDGs 1, 2, 3, 6, 11-15 compel us to shift the current agricultural practices towards sustainable farming. Therefore, farmers and policymakers must thoroughly quantify the impacts and explore available mitigation strategies to ensure responsible production and consumption. This chapter discusses the many models and toolkits that have been effectively used in agriculture and food systems to conduct a thorough evaluation of environmental impacts and serve as a foundation for making mitigation measures and framing policy decisions to achieve SDGs.

**Keywords:** agriculture, SDG, model, localization, LCA

## 2.1. Introduction

India, with a population of 1.3 billion as of 2015, is the seventh-largest country in the world by area. Its vast expanse encompasses a rich tapestry of climates, terrains, vegetation, fauna, and socioeconomic landscapes. India's land usage and cover have changed significantly throughout the past 140 years due to forest clearings, changes in agricultural methods, and increased urbanization. Remarkably, India ranks among the leading producers of agricultural commodities worldwide. In 2022-23, the share of GVA of agriculture and allied sector contributed a substantial 18.3% in the total economy and engaged approximately 45.76% of its workforce. Despite these economic contributions, India faces considerable challenges, most notably a relatively high poverty rate. Two-thirds of its population lives in rural areas, with a significant portion of individuals living below the poverty line defined by the World Bank, set at $1.90 per day (Hinz et al., 2020).

India's agricultural productivity has surged, yet persistent yield disparities (GYGA, 2018) remain, attributed to subsistence farming and limited access to inputs and technology (Balan et al., 2022). With its population set to surpass 1.6 billion by 2050 (Obiasi, 2017) and evolving dietary preferences, India faces the formidable task of scaling up food production significantly. Despite occupying merely 2.4 percent of the world's acreage, it currently sustains 18 percent of the global population, underscoring its critical responsibility in global food security amidst domestic challenges.

The vulnerability of the Indian agriculture sector to climate change is significant due to its heavy dependence on existing weather patterns, such as rainfall and temperature fluctuations, as well as water stress conditions common in major crop-producing states. This vulnerability is starkly obvious when combined with farmers' financial struggles. However, a pathway exists to mitigate these climate-related risks by bolstering farmers' ability to adapt and enhance the resilience of agricultural fields, thereby transitioning in the direction of sustainability (Kareemulla et al., 2017).

The core principles of sustainable agriculture (SA) systems are central to achieving this transition. These principles encompass various strategies, including the effective management of natural assets, the incorporation of localized and enhanced seed varieties, the amalgamation of conventional and innovative farming techniques, a decreased use of herbicides and pesticides, and a safeguarding of soil quality through the ethical utilization of bioresources (Rao et al., 2018). By embracing these key attributes, the agriculture sector can move toward a more secure and sustainable future and be better equipped to tackle the challenges posed by climate change.

Food and agriculture are critical (Selvaraj et al., 2021). Despite substantial advances in agricultural production to accomplish global food demand, it has presented tremendous damage to the ecosystem and the world at large (Srivastava et al., 2016; Rockstrom et al., 2017; Tilman et al., 2017; Ramankutty et al., 2018; Springmann et al., 2018). Water scarcity, soil degradation, ecosystem stress (Poore & Nemecek, 2018), biodiversity loss (Eyhorn et al., 2019), diminishing fish populations and forest areas, and increasing carbon emissions all contribute to the costs (Crippa et al., 2021). The efficiency of our natural resources has been harmed in several areas around the world, jeopardizing the planet's production efficiency in the long run. Even globally, the current state of the food system is unacceptable, with 815 million people hungry and one in three malnourished (FAO, 2019). This unbalanced food system is a threat to the world's population. Moreover, distress migration has reached an alarming level, one not seen in over 70 years (IOM, 2020). It is time we act decisively to preserve our natural resources and ensure a sustainable and equitable food system.

Poor performance in agriculture typically results in inflation, economic hardship, and a decrease in consumer spending, all of which can hinder the overall national economy. Moreover, agricultural growth in India tends to be more volatile than in countries like China. Therefore, India must elevate its growth trajectory while transitioning towards a more inclusive, stable, and resilient agricultural and agri-food system. To achieve this, there needs to be a shift in the ongoing agricultural debates, moving away from traditional methods solely focused on physical productivity targets. Instead, there should be a focus on developing a diversified, secondary, specialty, and resilient agriculture supported by intelligent policies, robust institutions, skilled human resources, and a conducive environment for both public and private sector investments and performance. Only then can the agricultural sector become resilient and sustainable, effectively tackling poverty, ensuring household nutritional security, and enhancing the livelihoods of millions of smallholder farmers in India.

Looking ahead, the 2030 Agenda for Global Goals clearly outlines the path to equitable prosperity. To deal with the enormous difficulties that lie ahead, the world must take revolutionary action, embrace sustainability, and address the underlying reasons for starvation and poverty.

Food and agriculture, as the vital connection between people and the environment, can aid in achieving various Sustainable Development Goals (SDGs). When properly nourished, young people can become capable, humanity can thrive and lead fulfilling lives, and society will progress. Crops, livestock, aquaculture, fisheries, and forests comprise the world's largest agriculture industry, providing most of the world's population with food and income. Sustainable food and agriculture have immense potential to enhance farming communities, create equitable expansion for nations, and accelerate progress toward the 2030 Agenda.

Policy formulation and execution, grounded in scientific rationale, hold significant sway in promoting the shift towards sustainable agriculture (SA) (Sternet et al., 2019). A pertinent illustration of this is evidenced by the thorough evaluation conducted by a team comprising scientists and agricultural experts in India. Their efforts culminated in developing the 'Report on Policies and Action Plan for a Secure and Sustainable Agriculture' (GOI, 2019), which underscores the imperative for SA systems in India and advocates for various policy adjustments to facilitate this transition. Establishing a structured procedure is essential for integrating sustainable food and agriculture into national development strategies and action plans, including building political momentum, a shared vision and action plan, and converting vision into action to accelerate change.

## 2.2. Genesis and Development of the SDGs

The SDGs for 2030 evolved from the Millennium Development Goals (MDGs) for 2015. The MDGs were a set of eight development goals established in the year 2000 with targets to be achieved by 2015. India's 2015 national report on the Millennium Development Goals highlight remaining challenges in reducing hunger, improving maternal health, enhancing access to water and sanitation, and addressing social and geographic disparities. Despite this, progress during the MDG era shows that with strong methods and concentrated efforts, global initiatives can be successful. India intends to carry this trend into the goals of the 2030 Agenda for Sustainable Development, ensuring inclusion. Achieving this involves awareness, human-centric approaches, policy portfolio, gender equality, technological advancements, finance models, transparency, and e-governance. Ultimately, sustainability hinges on shared human experiences, knowledge, decisions, and organizational efforts. Agriculture's sustainability relies on people's engagement and learning, with extension services playing a pivotal role in facilitating progress (Balan et al., 2023).

### 2.2.1. A Multidimensional Approach- 17 Goals and 169 Targets

The seventeen Sustainable Development Goals (SDGs), which include six social goals (Goals 1 through 6), five economic goals (Goals 7 through 11), and four environmental goals (Goals 12 to 15), are distributed and balanced among the three facets of sustainable development. A different goal (Goal-16) addresses development enablers, such as peaceful societies and efficient institutions. A stand-alone goal (Goal-17) on global partnership has been added to give developing nations systemic, technological, and financial support.

### 2.2.2. Institutional Set-Up in India

- *NITI Aayog:* The NITI Aayog is responsible for the national-level implementation of the Sustainable Development Goals (SDGs). As part of this effort, the NITI Aayog has conducted a comprehensive mapping exercise, identifying all SDGs, Central Ministries, and

Centrally sponsored Schemes. Additionally, it is engaging in consultations at both national and regional levels with various stakeholders, including States and Union Territories. Among its publications, the NITI Aayog has released the SDG India Index: Baseline Report 2018 (December 2018) and Localising SDGs: Early Lessons from India, 2019 (July 2019), alongside other documents. Based on the performance of the States/UTs, NITI Aayog has not only categorized them into Achievers, Front Runners, Performers and Aspirants but also has identified more than 100 aspirational districts for focused interventions.

- *MoSPI:* The Ministry of Statistics and Programme Implementation (MoSPI) is also a key player in implementing the SDGs. Given the significance of indicators in gauging progress and attainment of targets in India, MoSPI has formulated 306 national indicators aligned with the 169 SDG targets and the Global Indicators Framework. Among these, 62 priority indicators have been singled out to monitor the most critical developmental objectives in India.

### 2.2.3. A Global Indicator Framework: IAEG-SDGs

With 28 member countries, the UN Statistical Commission established the Inter-Agency and Expert Group on Sustainable Development Goals (IAEG-SDGs). India represents South Asia on IAEG-SDGs. According to the IAEG-SDGs' recommendation, the UN Statistical Commission developed a global indicator framework with 232 indicators. Indicators fall into three categories in the global framework: Tier 1, Tier II, and Tier III indicators (MoSPI, 2024).

- *Tier I:* The indicator has a well-defined conceptual framework, a globally recognized methodology, readily available standards, and regular data production from countries representing a minimum of 50% of the population in each region where the indicator is relevant.
- *Tier II:* Data is not consistently provided by nations, but the indicator is conceptually clear, has a globally recognized methodology, and available standards.
- *Tier III*: This indicator does not yet have an internationally recognized methodology or standard, but it is in the process of developing or testing one.

## 2.2.4. Progress Profile of India

According to the United Nations SDG Index and Dashboards Report 2023 (Sachs et al., 2023), which assesses countries' progress towards SDG, India ranks 112 out of 166 countries, with an overall index score of 63.5 percent (Figure 2.1) and a spillover score of 99.4 percent (Figure 2.3). Earlier, India's ranks were 121 in 2022, 120 in 2021, and 117 in 2020. For the second year in a row, the impact of the pandemic on SDG 1 and SDG 8 and its subpar performance on SDG 11–15 have caused the global average of the SDG Index to slightly decline in 2021. Massive humanitarian costs and military conflicts, such as the war in Ukraine, cause significant international repercussions on food security and energy prices. The climate and biodiversity crises compound these effects. It is also clear from Figure 2.2 that some of India's goals are progressing much faster than others, such as SDG 12, SDG 13, SDG 4, and SDG 1, in terms of reaching their targets. Limited progress is seen in the cases of SDG 8, SDG 7, SDG 6, SDG 3, SDG 14, SDG 17, and SDG 2. Other goals require more attention, and policy-level implementers must focus on these areas as well.

As of the 2021 report, India is on track to achieve the targets of SDG 12 and 13 well before 2030. As in the case of SDG 12, the achievement is maintained, but the pandemic has slightly affected climate action across the globe, and that is the reason for stagnating growth in the case of SDG 13. SDGs 2, 3, 5, 6, 9, 11, 14, 15, and 16 represent major challenges that the country faces, which vary from region to region. This clearly indicates that achieving targets with a one-size-fits-all approach is not feasible, and localizing the SDGs is essential.

The states and UTs of India were again studied to see the differences contributed towards achieving the target and based on their performance, they were categorized into 4 sets: aspirant, performer, front runner, and achiever. As clearly indicated (Figure 2.4), Chandigarh tops as the runner UT with the performance score of 79, and among states Kerala has the highest score of 75, followed by other states like Tamil Nadu, Himachal Pradesh, Andhra Pradesh, Uttarakhand, Goa, Karnataka, Sikkim, Maharashtra, Gujarat, Telangana, Punjab, Mizoram, Haryana and Tripura. Other states and UTs fall under the performer category, and if appropriate attention to improving indicators is taken, then the country's average score from 66 is likely to improve.

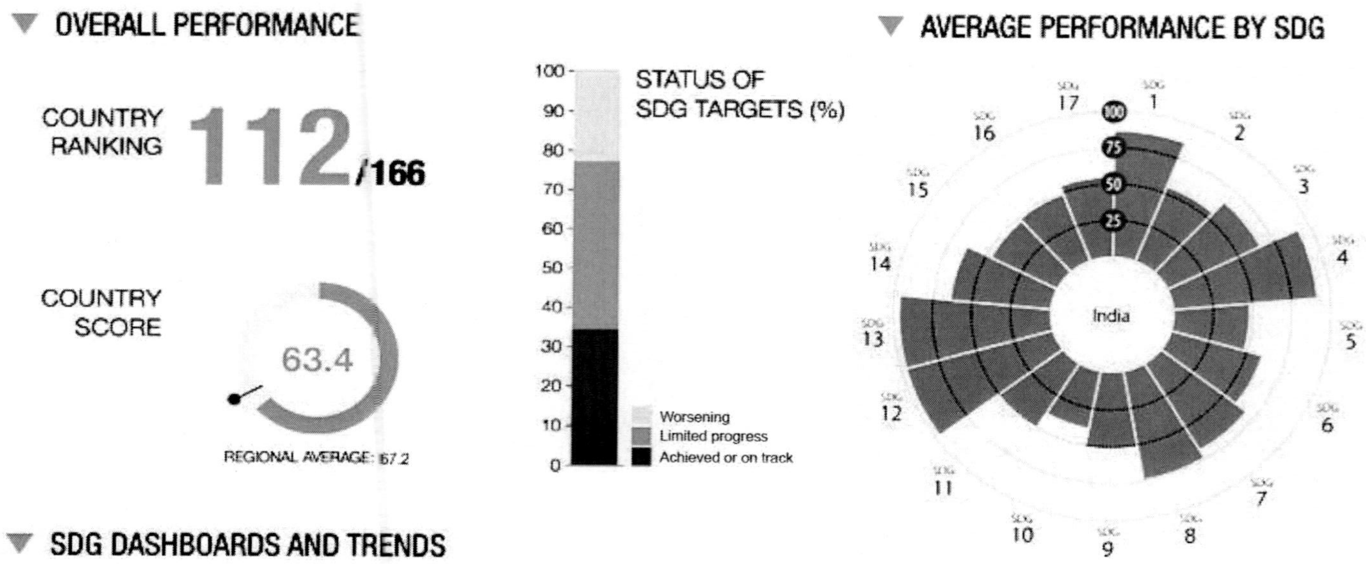


**Figure 2.1.** Sustainable Development Goals (SDGs).

**Figure 2.2.** Trends of SDG in India.

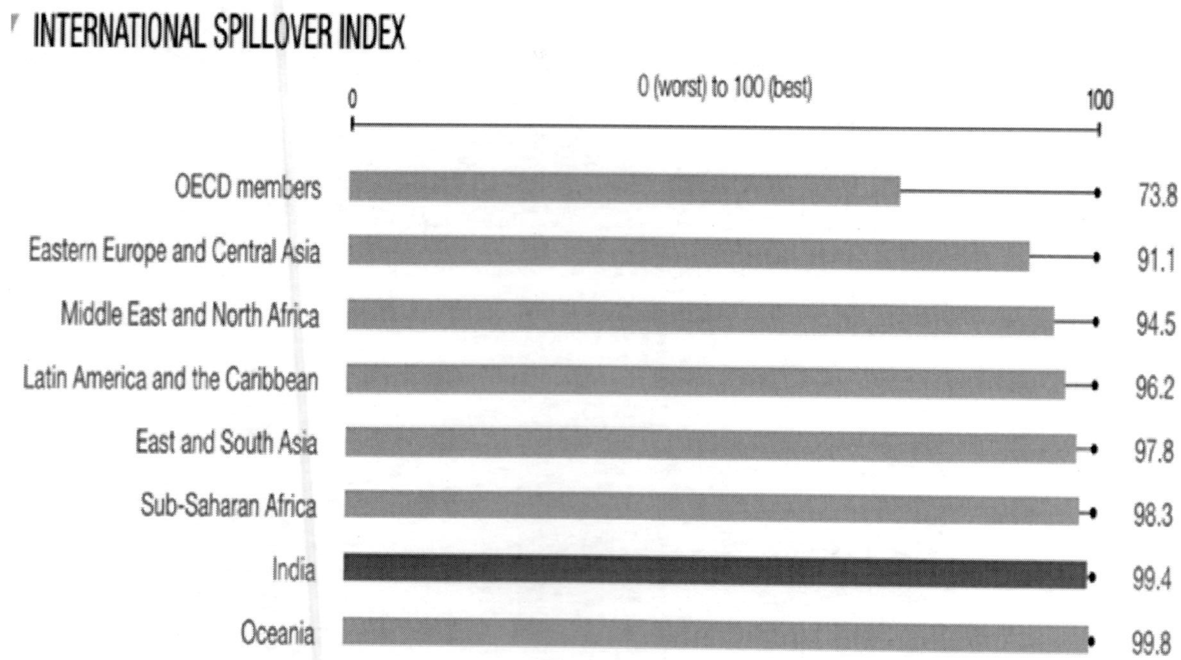


**Figure 2.3.** Comparison of spillover index with other neighboring countries.

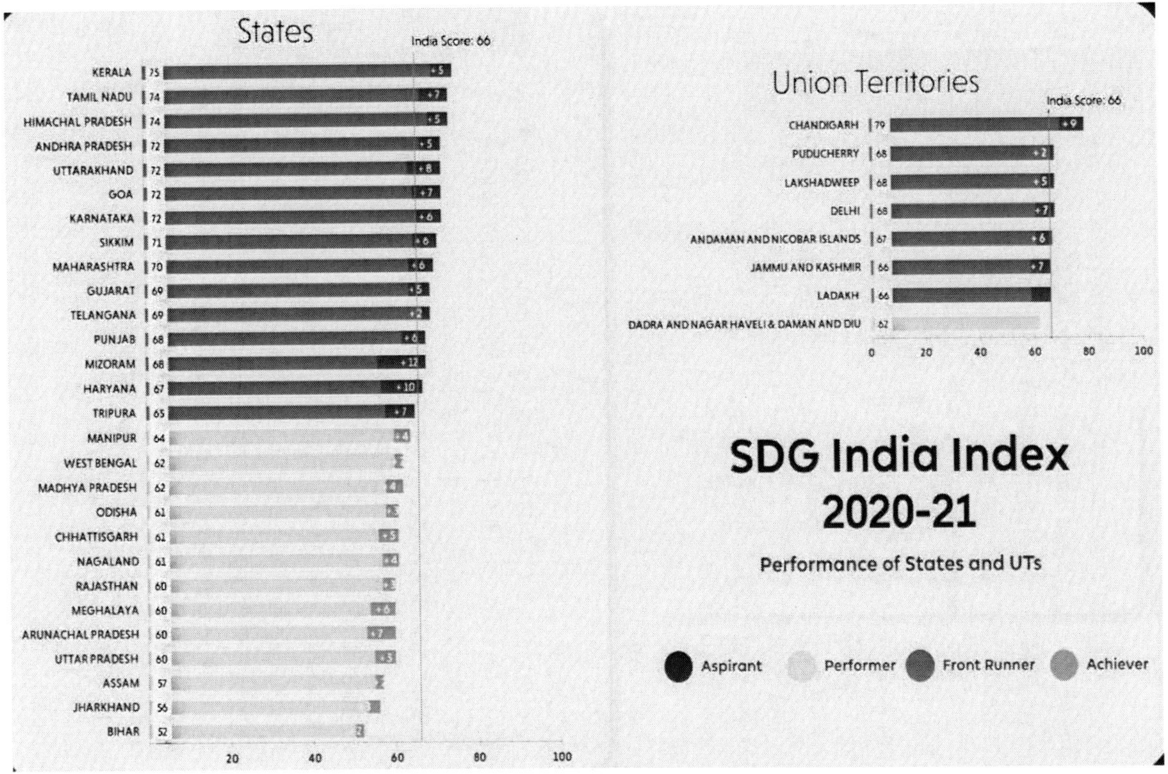


**Figure 2.4.** Localization to SDGs.

### 2.2.5. What Makes Agriculture Integral to Achieving Sustainable Development Goals?

#### *2.2.5.1. Agricultural Investments Stand Out for Their Unmatched Potential to Advance Food Security, Nutrition, Human Development, and Sustainable Economic Growth, Surpassing Other Sectors*

Relinquishment in agriculture is crucial for improving efficiency, productivity, and income growth as well as addressing poverty and hunger. According to the Sustainable Development Goals Report 2023, despite the pandemic prompting record-high nominal public expenditure on agriculture, reaching $700 billion in 2021, there was a notable decline in government spending in proportion with the sector's contribution to GDP. Except in Europe and North America, where historically significant government stimulus packages were enacted, the agriculture orientation index (AOI), which measures this trend, fell from 0.50 in 2015 to 0.45 in 2021 in the majority of regions. Notably, the AOI declined from 0.33 in 2015 to 0.21 in 2021 in Latin America and the Caribbean, marking the biggest decline in the region.

#### *2.2.5.2. Farmers in Developing Regions Can Attain Productivity Levels Comparable to Those in Developed Countries, Contributing to Sustained Progress in Sustainability Globally*

Between 2015 and 2021, there was a notable 14.6% surge in the total aid allocated to agriculture within developing nations, marking an increase from $12.8 billion to $14.2 billion when adjusted for base 2021 prices. Despite this significant growth in financial support, the global trade dynamics portrayed a contrasting picture for the Least Developed Countries (LDCs). In 2022, the proportion of exports originating from LDCs in the global merchandise trade remained disproportionately low, hovering at a mere 1.1 percent. This figure only saw a marginal uptick of 0.07 percentage points from the preceding year. Unfortunately, the goal of increasing the proportion of exports from Least Developed Countries by twofold by 2020, from its baseline value of 1.02 percent in 2011, remained unmet. Conversely, there was a notable expansion in the collective share of global merchandise trade for all developing nations. This share rose to 45.3 percent in 2022, showing a consistent upward trajectory over the past six years. This growth amounted to a substantial rise of 4 percentage points compared to the figures recorded in 2016.

### *2.2.5.3. Empowering Farmers Necessitates Enhanced Knowledge Eexchange and Accessible, High-Quality Extension Services for Effective Farm Management and Marketing*

Extension strategies underwent a gradual evolution in the wake of the emergence of farmer-first approaches within agricultural extension practices and the recognized inadequacies of the technology transfer model. This shift prioritized empowerment, active participation, capacity enhancement (particularly among farmer extension agents), and establishing or reinforcing local organizations as key objectives (Lubell et al., 2014). The agricultural market landscape is changing with the emergence of new market models like contract farming, farmers' markets, private wholesale markets, and internet retailers. By helping farmers to get better prices for their produce and so increase their incomes, these markets help to reduce poverty and inequality. Additionally, they facilitate cost reduction in production by fostering direct interaction between farmers and extension services (Paroda & Joshi, 2017). Information and Communication Technology (ICT) is vital for advancing agricultural and farming methods. To ensure fairness and prevent a digital gap, information concerning government policies, farmer-oriented programs, implementing institutions, new agricultural advancements, best practices, and sources of agricultural inputs (such as high-yield seeds and fertilizers) are communicated to farmers through information technology channels. The advantages of e-agriculture encompass access to pricing data, agricultural insights, both domestic and global markets, enhanced production efficiency, and establishing a supportive policy framework, all contributing to the betterment of farmers' livelihoods (Paroda, 2018).

### *2.2.5.4. Agriculture Relies on Supportive Frameworks Facilitating Investment in Infrastructure and Fostering Inclusive Markets*

Supply networks and agricultural markets are the development hubs. Markets for food and agriculture give customers more options and provide farmers with more incentives. Trade and well-functioning markets facilitate the best use of resources, spread information and technologies, and open doors for the network of supply chains for agricultural inputs and commodities that connect agriculture to other economic sectors. As a result, growth and development, as well as the economy's structural transformation, depend on the market system. Millions of people rely on markets to improve their standard of living, but they also have other advantages. For example, markets can help ensure food security by transferring food from places of surplus to areas of deficiency.

Both domestic and international markets must work well to propel economic growth. Markets might, nevertheless, occasionally be unable to balance the needs of the individual with those of society at large. Markets may fail to address social goals like eliminating inequality or having detrimental effects on the environment. Furthermore, inefficient and fragile agricultural and input supply chains make it difficult for them to quickly adapt to supply-side and demand-side shocks, endangering livelihoods and food security. Market failures and supply chain disruptions undermine the 2030 Agenda for Sustainable Development's core social, economic, and environmental benefits (FAO, 2022).

#### *2.2.5.5. With Appropriate Support, Farmers and Stakeholders Throughout the Value Chain Possess the Significant Capacity to Combat Food Waste and Loss*

All the parties involved in the coordinated production and value-adding processes required to create food products make up a food value chain or FVC. To be deemed sustainable, a food value chain must meet the following requirements: (i) it must be economically, socially, and environmentally sustainable; (ii) it must be cost-effective at every stage; and (iii) it must not negatively impact the environment. The SFVC concept acknowledges value chains as dynamic, market-driven systems, with specific value-added and sustainability, multidimensional performance metrics evaluated overall, and vertical coordination (governance) as the primary dimension. According to the SFVC development framework, farms and other agri-food operations behave and perform differently depending on a complex web of environmental, social, and economic settings, of which the FVC is the center. The value chain participants that manufacture or acquire goods from the upstream level modify those goods and then resell them to the next level make up the core FVC. These players do four tasks: distribution (wholesale and retail), aggregation, processing, and production. The aggregation stage is crucial for FVCs in developing nations because it can be difficult to efficiently aggregate and store tiny quantities of produce that are gathered from widely scattered smallholder farmers. A governance structure connects FVC participants to their larger operating environment. There are vertical links within the entire chain, such as when farmers enter into contracts with food firms to provide their produce, and lateral ties among the stakeholders at a specific chain step, such as when agri-producers organize into cooperatives. Value is ultimately decided by the food products people buy in domestic and international markets. It follows that SFVC development begins with a market opportunity and proceeds backward through the supply chain to determine what needs to be addressed to seize this opportunity. It is imperative for a developing country like India to track development and guarantee responsibility to pledges made and facilitate the achievement of transformation necessary to meet pre-decided targets. This is important in the phase of developing and revising national agricultural investment plans in the implementation of comprehensive agriculture programs. Finance proposals classify focus sections that would advance the sector's performance and SDGs. This chapter offers a collection of models and resources to help lead the attainment of the target and quantitatively evaluate the activities taken to achieve the SDGs, particularly of agriculture. This will facilitate the transition to evidence-based, sustainable food and agriculture more quickly.

## 2.3. Agriculture and the SDGs: The Interconnected Linkage

### 2.3.1. Goal 1 - No Poverty

The study 'Localising multidimensional poverty assessments for inclusive public policies: The case for a communal poverty profile in Mali' by Missinhoun (2020) reveals that reliable and up-to-date data beyond national averages to uncover geographic and other inequalities is one of the major challenges for monitoring progress towards SDG. In this context, a typology of communes (people in the entity) was generated according to their multidimensional poverty status. The results indicated that nearly half of the 703 communes (49.5 percent) are poor in monetary terms, and 26 percent face food poverty. Food poverty increases with a decline in land's agricultural potential. In terms of nutrition, 64.1 percent combine high level of chronic malnutrition with a moderate level of acute global malnutrition, and 35.9 percent combine a high level of acute global malnutrition with a moderate level of chronic global malnutrition. The first group is more associated with pastoral areas, while the second is associated with transhumance and nomadic livestock areas. Child malnutrition is chronic in zones with soil having agricultural potential and acute in zones with soil unsuitable for agriculture.

### 2.3.2. Goal 2 - Zero Hunger

The Food Insecurity Experience Scale (FIES) has the potential to capture the complexity of actual and perceived food security (FAO, 2019). A case study analysis of Nigeria (Gil et al., 2019) reveals that the performance of Nigeria's agriculture is tied to macro-development issues like land fragmentation (which increases transaction costs and limits mechanization), high vulnerability to climate shocks (farmland is mostly rainfed), and weak agriculture services (limited infrastructure, technical assistance, access to credit, and access to fertilizers). Despite extensive global initiatives, as of 2022, approximately 45 million children under the age of 5 were afflicted by wasting, 148 million suffered from stunted growth, and 37 million grappled with obesity conditions. A substantial reorientation of efforts is imperative to attain the nutrition goals set for 2030. Coordinated efforts are required to end hunger by 2030, including educating people about nutrient-rich foods like millet and pulses, rewarding their production, and promoting farmhouse gardening for personal consumption. This requires coordinated action, policy reforms, investment in sustainable agriculture, and mitigating the impacts of conflicts and pandemics on global nutrition and food security.

### 2.3.3. Goal 3 - Good Health and Well-Being

Good health begins with nutritious food, which provides energy and essential nutrients for growth, development, and active lifestyles. Agriculture is key to disease prevention and health enhancement, empowering families to produce, purchase, and consume more food. Enhanced crop varieties rich in essential vitamins improve people's nutritional intake, while increased farmer incomes enable access to various nutritious foods. A country's ability to produce food in the future depends on its ability to diversify its crop mix and preserve its natural resources, which are the cornerstones of all agricultural and food systems. Educational initiatives focused on nutrition, involving collaboration between agricultural and healthcare professionals, especially targeting families with young children, have the potential to enhance household dietary practices substantially. Encouraging the use of locally cultivated, healthy food supports the growth of the community's economy.

### 2.3.4. Goal 6 - Clean Water and Sanitation

As the population continues to grow in the future, unless effective mitigation strategies are implemented, the escalating rate of per capita total water usage will exacerbate water scarcity. Some potential strategies include enhancing agricultural water productivity and irrigation efficiency, reducing domestic and industrial water consumption, controlling population growth (considered "soft measures"), as well as boosting water storage in reservoirs, and expanding seawater desalination (considered "hard measures") (Roy and Pramanick., 2019; Wada et al., 2014). According to the Sustainable Development Goals Report (2023), around 2.4 billion individuals resided in countries facing water stress in 2020, with nearly 800 million living in nations categorized as having high to critically high levels of water stress. Water-use efficiency worldwide rose 9 percent, from \$17.4/m$^3$ in 2015 to \$18.9/m$^3$ in 2020. It ranges from below \$3/m3 in economies that depend on agriculture to over \$50/m$^3$ in highly industrialized or service-based economies. Agriculture experienced the greatest increase in water-use efficiency (20%) from 2015, compared with the industrial and service sectors (13 and 0.3 %, respectively). Optimizing industrial and energy cooling processes, addressing leaks in distribution networks, managing agriculture more effectively, and increasing irrigation efficiency are all necessary to increase water use efficiency. Safeguarding aquatic ecosystems, maintaining good sanitation and hygiene standards, managing water resources sustainably, and providing clean drinking water are all necessary to create a world where everyone has consistent access to it. A transparent and effective governance framework must be implemented to distribute water resources fairly across different sectors, such as agriculture, industry, and urban regions (FAO, 2018).

### 2.3.5. Goal 11 - Sustainable Cities and Communities

Agriculture is acknowledged for its pivotal role in bolstering urban environments' economic, social, and ecological sustainability. Consequently, municipal authorities can endorse urban agriculture as a strategy to realize sustainable development aims. The study by Alves et al., (2024) conducted in Port Alegre, Brazil indicates that urban agriculture in the city manifests itself in economic, social, and environmental terms. In 2019, town areas in Eastern and Southeast Asia, home to a large portion of the global population, faced worse air quality than cities did. Addressing air pollution necessitates a shift of perspective, acknowledging that it extends beyond urban areas. Cities serve as fertile grounds for urban agriculture, catering to proponents of compact city designs that emphasize energy management and pollution control (Azunre et al., 2019). Local markets are crucial for sustaining urban agriculture and ensuring steady income for farmers and their families, along with residents in nearby urban areas. This fosters their continued engagement in agricultural pursuits, promoting social equity and community cohesion by bringing together people of diverse backgrounds, ages, and cultures (Curan & Marques, 2021). Urban agriculture can have a positive impact on household food security by fostering the creation of direct livelihoods, especially in impoverished communities.

### 2.3.6. Goal 12 - Responsible Consumption and Production

According to "Sustainable Development Goals Report 2023," 13.2 percent of the world's food was lost after harvest along the supply chain from farm to consumer as of 2021, even though 828 million people were hungry. This percentage has barely changed since 2016 and falls short of the aim of significantly reducing post-harvest food losses by 2030. An extra 17 percent of food is wasted at the household, food service, and retail levels, culminating in a staggering 931 million tons of food waste, or 120 kg per person, in 2019.

### 2.3.7. Goal 13 - Climate Action

Agriculture, forestry, and land use sectors contribute to 18.4 percent of global greenhouse gas emissions. Transition to low-carbon and resource efficient energy production systems can partially address this issue. The total global carbon emission is estimated to be 52 million tonnes ($CO_2$ equivalent) of which the contribution from global food production is 26 per cent. Out of this, 26 percent is within the food production system, 27 percent is from crop production, 31 percent is from livestock and fisheries, 24 percent from land use change and management, and the rest, 18 percent is from the supply chain, including transport, packaging and food processing (Poore & Nemeck, 2018). Climate change intensifies food insecurity, with projected arable land reductions in South America (1-21%), Africa (1-18%), Europe (11-17%), and India (20-40%). In a warmer and higher $CO_2$-world, changes in food quality like decreased protein, mineral nutrient concentration, and altered lipid composition are also likely to bother (Abraham & Pingali, 2020; Rivero et al., 2021). Conservation technologies like zero tillage, direct seeded rice, and laser land leveling can mitigate climate change risks, reduce GHG emissions, and enhance carbon sequestration. In India, the National Bank for Agriculture and Rural Development (NABARD) implements agriculture and rural development programs, including Farmer Producer Organizations (FPOs)/ self-help groups, Agri-infrastructure development, value chain enhancement, and climate change management. Policy actions for climate risk management include investing in climate-smart technologies, leveraging adaptation and mitigation co-benefits, integrating land-use policy with crop planning, and promoting climate education while investing in capacity building.

### 2.3.8. Goal 14 - Life Below Water

In today's world, the significance of maintaining the health of our oceans and seas cannot be overstated. Encompassing over 70 percent of the Earth's surface, these bodies of water play a vital role by supplying half of the planet's oxygen, capturing carbon, and serving as habitats for 80 percent of Earth's biodiversity. Farming, aquaculture, and wastewater management are adding to nutrient levels in coastal regions, leading to inescapable eutrophication and the proliferation of algal blooms. These blooms result in oxygen depletion, endanger marine organisms, pollute seafood, and harm seagrass and coral reefs, among other effects. The repercussions are significant for marine ecosystem well-being, local communities, fisheries, and tourism. As per the SDG report for 2023, satellite data indicates heightened coastal eutrophication trends worldwide in 2022 compared to the baseline of 2000-2004, albeit varying in magnitude from recent periods. Persistent elevated rates were observed in the Arabian Sea, where algal blooms posed risks to fisheries, tourism, and desalination facilities, notably during late winter and early spring, spanning from 2018 to 2022.

### 2.3.9. Goal 15 – Life on Land

According to "Sustainable Development Goals Report 2023", terrestrial ecosystems are vital for sustaining human life, contributing to over half of global GDP and encompassing diverse cultural, spiritual, and economic values. Forests are our planet's major carbon and biodiversity repositories, critical for combating climate change and sustaining essential goods, services, and livelihoods. Despite conservation efforts, nearly 100 million hectares of net forest area have been depleted over the past two decades. Between 2000 and 2020, global forest coverage reduced from 31.9 percent (4.2 billion hectares) to 31.2 percent (4.1 billion hectares). Nearly 90 percent of global deforestation is directly attributed to agricultural expansion, with cropland accounting for 49.6 percent and livestock grazing for 38.5 percent. Oil palm harvesting alone was responsible for 7 percent of global deforestation from 2000 to 2018. Agriculture expanded into forests in Latin America, the Caribbean, sub-Saharan Africa, and Southeast Asia from 2015 to 2020. Conversely, Asia, Europe, and North America saw forest maintenance or growth. It is crucial to globally and regionally sustain forests for their ecological, social, and economic roles, especially in developing countries and the tropics.

## 2.4. Ways for Improving Quantitative Assessment of Sustainable Agriculture

### 2.4.1. Synergies and Trade-Offs

In recent years, extensive research has delved into understanding how the SDGs interact, exploring both their trade-offs and synergies. This research investigates how individual SDGs interact either with other goals or on a global scale. Methodologically, it utilizes official SDG indicator data for quantitative analysis, expert input, and qualitative frameworks. Generally, more synergies—where progress in one goal contributes to another—than drawbacks have been identified. Anderson (2021) highlights the critical importance of second-order effects in SDG development and system modeling. This involves a thorough correlational analysis of all possible pairs of SDG indicators, identifying target pairs that exhibit global representativeness and strong directional evidence of synergy or trade-off. Subsequently, these pairs are extracted for further analysis, including regional and sensitivity analyses using the iMODELER software. Survey findings from 22 nations, including China, India, Korea, and Peru, reveal that over 30% of trade-offs were observed. Notably, efforts to enhance resource efficiency in production and consumption may face challenges due to the formalization and expansion of Micro, Small, and Medium Enterprises (MSMEs), rather than a shift towards non-agricultural employment. Conversely, informal economies demonstrate potential advantages over formal ones, such as greater resource efficiency, creativity, low-carbon practices, and resilience to climate change.

### 2.4.2. Integrating Modeling Framework

The utilization of scientific methods to guide policy toward achieving the SDGs often involves sector-specific evaluations, which frequently overlook the interconnectedness highlighted in multisectoral assessments. Hinz (2020) suggests that integrating multiple spatially explicit models with an economic model is appropriate. In the study, this integrated modeling framework includes Land SHIFT (Schaldach et al., 2011), the International Model for Policy Analysis of Agricultural Commodities and Trade (IMPACT; Robinson et al., 2015), and two empirical models for analyzing land use change effects on biodiversity (Alkemade et al., 2013; Jenkins et al., 2013), along with changes in carbon stock in soils and vegetation (European Commission, 2010).

The findings indicate that almost all predicted crops exhibit low yields due to climate change, with temperate tubers and roots being notable exceptions, benefiting from the shifting climate. The growth rate of spatially aggregated yields of all crops under the climate change scenario (+41%) from 2010 to 2030 is lower than that under the no climate change scenario (+48%). Similar trends are observed in examining the growth rates of total crop production, with climate change leading to lower production (+43%) compared to a scenario without climate change (+50%). These scenarios also suggest shifts in the importance of various crop groups, with some crops gaining prominence while others receive less cultivation. Commercial crops like cotton, wheat, and vegetables exhibit the highest productivity and yield increases.

More recently, Yue et al. (2021) proposed an integrated modeling framework for the sustainable agriculture energy-water-food nexus (EWFN), employing an inexact multi-objective optimization approach. This framework incorporates methods such as life cycle assessment (Al-Ansari et al., 2015), input-output modelling (Deng et al., 2020), footprint assessment (Lu et al., 2021), econometric modeling (Ozturk, 2015), and system dynamic modeling (Naderi et al., 2021). This decision framework offers ecosystem-based solutions for optimal water and land resource allocation under different scenarios towards sustainability.

The primary objective of this model is to maximize the social welfare of water allocation, hydroelectric generation, grain production, and positive farmland service value while minimizing negative farmland service value. This directly contributes to SDGs 2 (food security), 6 (water security), 7 (energy security), 8 (economic growth), and 13 (climate change mitigation), while also indirectly supporting SDGs 3 (human well-being), 11 (sustainable cities), 12 (sustainable production), and 15 (sustainable land ecosystems).

### 2.4.3. Composite Indicators

Paoli and Addeo (2019) reported in their study that composite index development is a method to measure sustainability where two-stage PCA was used, and then SDGs were aggregated into three pillars, namely, social, economic, and environmental. Goals 1, 3, 4, 5, 11, and 16 were under the social category; the economic set consisted of goals 8, 9, 10, and 17, and the others under Environment, namely goals 2, 7, 12, 13, and 15. This helps to indicate member states' performance into 5 categories, namely, worst, negative, medium, good, and top performer. There are pros and cons with composite indicators. Composite indicators abridge complex multidimensional issues that can be used to support decision-making. Aggregation helps with a more straightforward interpretation and ranking of countries on complex issues. They can be used to communicate issues of public interest by providing a summary of the performance of one country compared to others, including tracking performance over time. It can help reduce the size of the list of indicators to include more information with a limited size. Composite indicators may offer inaccurate and unstable policy directives when constructed inadequately or misinterpreted. Responsiveness/changeability analysis can be done to measure the robustness to avoid such situations.

### 2.4.4. Life Cycle Assessment

In an age marked by heightened environmental awareness and an imperative for sustainable practices, it becomes paramount to grasp agricultural value chains' complete life cycle impact. Life Cycle Assessment (LCA) emerges as a mode utilized to appraise the environmental ramifications caused by a product, mechanism, or system, from procurement of inputs to discarding.

This method adopts a thorough and structured approach, encompassing all life cycle phases, including production, utilization, and final stages (VoglhuberSlavinsky et al., 2022).

LCA furnishes a holistic environmental impact perspective by scrutinizing diverse marks like carbon emissions, power consumption, water consumption, land occupancy, and toxicity potential. It considers the resources consumed, emissions discharged, and waste generated at each stage, facilitating a comprehensive understanding of environmental aspects.

Ensuring responsible production and consumption is imperative to arrive at the ambitious sustainable development goals outlined by the UN. Environment evaluation within a globally recognized life cycle assessment framework is indispensable to avert any burden shifting. Nevertheless, such analyses often focus solely on specific environmental aspects, such as estimating greenhouse gas emissions. Furthermore, social life cycle assessment, including agriculture, remains nascent across industrial sectors (Reinales et al., 2020). Consequently, no comprehensive sustainability approach exists for agriculture and food systems, nor are there guidelines for developers. Given the increasing demand for life-cycle assessment in projects by public authorities and the public's call for mandatory life-cycle assessments like carbon and water footprints for commercial farms in the foreseeable future, the urgency for such guidelines is evident.

The overarching goal of developing appropriate models and statistically quantifying data is to support efforts to achieve sustainable development goals. By employing rigorous research methodologies and statistical techniques, stakeholders can generate actionable insights, identify effective solutions, and measure progress toward sustainable development objectives. Whether it is addressing poverty, promoting inclusive economic growth, combating climate change, or ensuring environmental sustainability, the systematic application of appropriate models and statistical quantification plays a pivotal role in guiding interventions, mobilizing resources, and fostering collaboration across sectors and stakeholders. A critical assessment of a few models with spatial extent up to Asia has been included in Table 2.1.

**Table 2.1.** Modeling for sustainable development goals

| Model | Particulars | Purpose | Policy area | How the model can contribute | SDG | Spatial extent |
|---|---|---|---|---|---|---|
| AGLINK-COSIMO | Bottom-up partial equilibrium (PE) model, agricultural commodity market model, dynamic, deterministic, stochastic temporal extent-very short term (<1 yr), short-term (period of 5 years or less), medium-term | Short-term forecasting, medium to long-term exploring (baselines), or scenario analysis | Agriculture and rural development | Economic impact assessment | 2, 12 | Europe, America, Africa, Asia, Oceania |
| CAPRI | Bottom-up PE model, bottom-up optimization model, static, deterministic modelling | Medium to long-term exploring (baselines) or scenario analysis, Policy impact assessment | Agriculture and rural development, climate action, environment, regional policy | Model simulates policy impacts on GHG emissions, ammonia emissions, nitrates leaching, soil erosion, C sequestration, trade impact analysis, simulate impacts of mitigation policies (from CO2 taxes to subsidies) | 2, 6, 12, 13, 15, 17 | EU member states, SAARC, BRICS, ASEAN, OECD, Arab common market, OPEC, ALADI, Andean community, CACM, EFTA, CAIS, LAES, CAEMC, WAEMU countries |

| Model | Particulars | Purpose | Policy area | How the model can contribute | SDG | Spatial extent |
|---|---|---|---|---|---|---|
| EFFIS | Top-down input-output model, bottom-up multi agent model, bottom-up simulation model, statistical model, model solution- simulation, monitoring and forecasting, dynamic, deterministic and probabilistic modeling approach | Short-term forecasting, medium to long-term exploring or scenario analysis, medium to long-term backcasting | Agriculture and rural development, international cooperation and development, environment, regional policy, humanitarian aid and civil protection | Building more resilient society, adapting technologies and developing risk reduction strategies to prepare for climate change, analysis of wildfire regimes, assessment of wildfire impact including deforestation, degradation and desertification | 1, 3, 11, 13, 15, 17 | All European countries, America, Africa, Asia and Oceania |
| MAGNET | Top-down computable general equilibrium model, simulation, static, dynamic, deterministic | Medium to long-term exploring or scenario analysis | Agriculture and rural development, environment, economy, finance, energy, regional policy | A global whole-economy model used to analyze policy scenarios on agricultural economics, bioeconomy, food security, and international trade. E.g., indicators-trade and migration scenarios | 1, 2, 7, 8, 9, 10, 12, 13, 15, 17 | EU member states, Andean community, Arab common market, CACM, ASEAN, EFTA, OECD, OPEC, Western Balkan, Benelux, ACP, Comecon, Caricom, Ecowas, Mercosur, Mediterranean, Central and Eastern European, APEC, FTA, CEFTA, BRICS countries |

**Table 2.1**. (Continued)

| Model | Particulars | Purpose | Policy area | How the model can contribute | SDG | Spatial extent |
|---|---|---|---|---|---|---|
| POLES | Bottom-up PE model, simulation, dynamic, deterministic | Medium to long-term exploring or scenario analysis | Agriculture and rural development, environment, climate action, energy, transport | It is a world-wide energy demand and supply and greenhouse gas emissions model used for international energy and climate policy assessment | 1, 2, 3, 6, 7, 8, 9, 11, 12, 13, 15, 17 | All countries of Europe, America, Africa, Asia, Oceania, EU member states, SAARC, BRICS, ASEAN, OECD, Arab common market, OPEC, ALADI, Andean community, CACM, EFTA, CAIS, LAES, CAEMC, WAEMU countries |
| TM5-FASST | Bottom-up simulation model reduced from process-based model, simulation, static, deterministic | Medium to long-term exploring or scenario analysis, medium to long-term backcasting | Agriculture and rural development, environment, climate action | Make swift air quality-related impact assessments (health, crop yield, near-term climate) of past, present and future pollutant emission scenarios on a global to regional scale | 3, 11, 13, 15 | Global |

Source: JRC, 2019.

## 2.5. Discussion

Various models and approaches exist to quantify progress towards achieving SDGs. However, using the right model that best fits the country's situation is important for making relevant decisions. Models are crucial for the following reasons:

1. Measuring progress- SDGs address global challenges such as poverty, inequality, climate change, and environmental degradation. Models help quantify progress toward goals, enabling governments, organizations, and policymakers to track their progress toward the target.
2. Accountability—Models in place provide a framework for holding stakeholders accountable for their commitments to the SDGs.
3. Resource allocation models help identify priorities and allocate resources effectively. By understanding which areas are lagging or most need attention, governments and organizations can allocate funding and resources where they are most needed to maximize impact.

Current reporting rates support the critical requirement for national capacity building to guarantee that the majority of countries can generate and report on the majority of SDG indicators by 2030. Although obstacles differ among nations, current evaluations point to the following as the primary barriers to reaching the SDGs:

1. National and regional metrics are not in line with the Global Indicator Framework
2. Inadequate or ineffective spending on SDG data generation and application by custodian organizations and national governments
3. Inadequate national SDG coordination, especially between government agencies and ministries
4. Inadequate capacity to gather, aggregate, summarize, and apply SDG indicators
5. The UN system's agencies do not sufficiently coordinate with one another on SDG data (e.g., in planning and supporting integrated data collecting and capacity development efforts)
6. National financial resource limitations

7. Countries' enduring misunderstanding of the 2030 Agenda's politics and data
8. The custodian agencies' insufficient investment money for initiatives supporting SDG monitoring and follow-up
9. Inconsistent standards for SDG indicator applicability in various nations result in poor indicator uptake or selective indicator use.

The food industry, agriculture, and the land used for farming extensively impact the nation's economy and farmers' well-being. A quantitative analysis of agrarian maintainability elucidates varied needs among countries, indicating the need for SDG localization. This supports the reports of Zhang et al., 2021. The current policies, such as NMSA, include several SA-focused initiatives, although they do not address the total agriculture value chain. In accordance with the study of Priyadarshini and Abhilash (2020), the formulation of a comprehensive 'National Policy on Sustainable Eco-Agri-Food Systems' might act as a framework for integrating sustainability ideas throughout the whole value chain of agriculture. Social capital, encompassing networks, relationships, and norms that promote cooperation and collective endeavors, has been instrumental in India's attainment of the Sustainable Development Goals. It enabled communities to share resources, work towards common goals, and facilitate development program implementation, promoting inclusive decision-making processes. Promoting social cohesion, mobilizing resources, and fostering partnerships are essential for sustainable development, and by leveraging social capital, India can further harness the collective efforts of individuals, communities, and organizations to various challenges and achieve SDGs. They do not address the full agriculture value chain. In accordance with the study of Priyadarshini and Abhilash (2020), the formulation of a comprehensive 'National Policy on Sustainable Eco-Agri-Food Systems' might act as a framework for integrating sustainability ideas throughout the whole value chain of agriculture. Social capital, encompassing networks, relationships, and norms that promote cooperation and collective endeavours, has been instrumental in India's attainment of the Sustainable Development Goals. It enabled communities to share resources, work towards common goals, and facilitate development program implementation, promoting inclusive decision-making processes. Promoting social cohesion, mobilizing resources, and fostering partnerships are essential for sustainable development, and by leveraging social capital, India can further harness the collective efforts of individuals, communities, and organizations to various challenges and achieve SDGs.

In short, agricultural sustainability is multifaceted, with varying perspectives on its dimensions. Developing policy analysis frameworks for SDG localization can enhance transformation potential, impacting both rural and urban sustainability. However, effective implementation hinges on access to sufficient data and resources, which are not uniformly available across state governments. By engaging state governments with global sustainability frameworks, this localization approach has the potential to accelerate progress towards SDGs significantly.

## Conclusion

Agricultural sustainability is central to achieving the Sustainable Development Goals (SDGs), given its direct and indirect impacts on economic stability, environmental health, and societal well-being. The integration of quantitative models and assessment tools enables policymakers, researchers, and stakeholders to measure progress, identify gaps, and implement targeted interventions. However, the effectiveness of these approaches depends on factors such as data availability, coordination among institutions, and resource allocation. A key challenge in SDG implementation is the misalignment of national and regional metrics with the Global Indicator Framework, alongside insufficient financial and institutional support. Strengthening national capacities, fostering cross-sectoral coordination, and enhancing policy frameworks—such as the proposed 'National Policy on Sustainable Eco-Agri-Food Systems'—can significantly contribute to the localization of SDGs in agriculture. Furthermore, leveraging social capital through community networks and collaborative governance can amplify the impact of sustainability initiatives. To accelerate progress toward SDGs, it is crucial to integrate sustainability principles across the entire agricultural value chain, ensuring responsible production and consumption. With strategic planning, improved data management, and inclusive policymaking, agricultural systems can become more resilient, equitable, and environmentally sustainable, ultimately driving long-term transformation in the food and farming sectors.

## References

Abraham, M., Pingali, P. L. (2020). *Climate change and food security. In Book 'Population, agriculture, and biodiversity: Problems and prospects.* 97-122. University of

Missouri press. HTTPs://www.researchgate.net/publication/361824139_CLIMATE_CHANGE_AND_FOOD_SECURITY.

Alkemade, R., Reid, R. S., van den Berg, M., de Leeuw, J., & Jeuken, M. (2013). Assessing the impacts of livestock production on biodiversity in rangeland ecosystems. *Proceedings of the National Academy of Sciences of the United States of America*, 110(52), 20900–20905. https://doi.org/10.1073/pnas.1011013108.

Al-Ansari, T., Korre, A., Nie, Z., Shah, N. (2015). Development of a life cycle assessment tool for the assessment of food production systems within the energy, water and food nexus. *Sustain. Prod. Consum.* 2, 52–66.

Anderson, C. C., Denich, M., Warchold, A., Kropp, J. P., and Pradhan, P. (2021). A system model of SDG target influence on the 2030 Agenda for Sustainable Development. *Sustainability Science*, 17: 1459-1472. https://doi.org/10.1007/s11625-021-01040-8.

Azunre, G. A., Amponsah, O., Peprah, C., Takyi, S. A., & Braimah, I. (2019). A review of the role of urban agriculture in the sustainable city discourse. *Cities*, 93, 104-119. https://doi.org/10.1016/j.cities.2019.04.006https://doi.org/10.1016/j.cities.2019.04.006.

Balan, A. S., Indulekha, V. P., Safiya, N. E., Surendran, D., Ashitha, M. R., Rani, D. C. V. (2023). Technology needs assessment of cool season vegetable growers in Kerala. *The Pharma Innovation Journal* Available online :https://www.thepharmajournal.com/archives/2023/vol12issue2/PartJ/11-12-858-109.pdf.

Balan, A. S., Ngasainao, C., and Seema, B. (2023). "Extension in Sustainable Development Goals: Contributing to Global Agenda." In Book *Advances in Agriculture Extension* Vol 1, edited by Pooja Jena, Dinesh Suryawanshi, Vishwanath H, Kalyani Konch, and Radhika, 292-311. New Delhi: Elite Publishing House.

Crippa, M., Solazzo, E., Guizzardi, D., Monforti-Ferrario, F., Tubiello, F. N., Leip, A. (2021) Food systems are responsible for a third of global anthropogenic GHG emissions. *Nat Food* 2:198–209. https:// doi. org/10.1038/s43016-021-00225-9.

Curan, R. M., & Marques, P. E. M. (2021). Multifuncionalidade da agriculturaurbana e periurbana: umarevisãosistemática. EstudosAvançados, 35, 209- 224. https://doi.org/10.1590/s0103-4014.2021.35101.013.

Deng, H. M, Wang, C., Cai, W. J., Liu, Y., Zhang, L. X. (2020). Managing the water-energy- food nexus in China by adjusting critical final demands and supply chains: an input- output analysis. *Sci. Total Environ.* 720, 137635.

European Commission (2010). Commission decision of 10 June 2010 on guidelines for the calculation of land carbon stocks for the purpose of annex V to Directive 2009/28/EC. OJ,151 https://eurlex.europa.eu/LexUriServ/LexUriServ.do?uri=OJ:L:2010:151:0019:0041:EN:PDF.

Eyhorn, F., Mueller, A., Reganold, J. P., Frison, E., Herren, H. R., Luttikholt, L., Mueller, A., Sanders, J., Scialabba, N. E. H., Seufert, V., Smith, P. (2019). Sustainability in Global agriculture driven by organic farming. *Nat. Sustain.* 2, 253–255. https://doi.org/10. 1038/s41893-019-0266-6.

FAO (2018). Transforming food and agriculture to achieve the SDGs, 20 interconnected actions to guide decision makers.

FAO (2019). FAO and the SDGs Indicators: Measuring up to the 2030 Agenda for sustainable development. Available at: https://www.fao.org/3/i6919e/i6919e.pdf.

Gil, J. D. B., Reidsma, P., Giller, K., Todman, L., Whitmore, A., Ittersum, M, V. 2019. Sustainable development goal 2: Improved targets and indicators for agriculture and food security. *Ambio*. 48:685-698.https://doi.org/10.1007/s13280-018-1101-4.

GOI (2019). Report on Policies and Action Plan for a Secure and Sustainable Agriculture, Submitted to Principal Scientific Advisor. Available at:https://psa.gov.in/reports.

GYGA (2018). Global yield gap atlas. Wageningen University and Research, University of Nebraska—Water for Food. http://www. yieldgap.org/web/guest/home. Accessed Mar 2024.

Hinz, R., Sulser, T. B., Huefner, R., Mason-D'Croz, D., Dunston, S., Nautiyal, S., Ringler, C., Schuengel, J., Tikhile, P., Wimmer, F., Schaldach, R. (2020). Agricultural development and land use change in India: A scenario analysis of trade-offs between UN Sustainable Development Goals (SDGs). *Earths future*, 8(2) https://doi.org/10.1029/2019EF001287.

IOM (2020). IOM UN Migration, World migration report 2020. Available at: www.iom.int/wmr.

Jenkins, C. N., Pimm, S. L., & Joppa, L. N. (2013). Global patterns of terrestrial vertebrate diversity and conservation. *Proceedings of the National Academy of Sciences of the United States of America*, 110(28), E2602–E2610. https://doi.org/10.1073/PNAS.1302251110.

JRC (2019). JRC Statistical audit of the sustainable development goals index and dashboards. Available at: https://publications.jrc.ec.europa.eu/repository/handle/JRC116857.

Kareemulla, K., Venkattakumar, R., Manoj, P., Samuel.(2017). An analysis on agricultural sustainability in India. *Curr. Sci.* 112 (2), 258–266. https://doi.org/10.18520/cs/v112/i02/258-266.

Lubell M, Niles M, Hoffman M (2014) Extension 3.0: managing agricultural knowledge systems in the network age. *Soc Nat Resour* 27(10):1089–1103.

Lu, S. B., Zhang, X. L., Peng, H. R., Skitmore, M., Bai, X., Zheng, Z. H., 2021. The energy-food-water nexus: water footprint of Henan-Hubei-Hunan in China. Renew. *Sustain. Energy Rev.* 135, 110417.

MoSPI (2024). Global indicator framework: IAEG-SDGs. Available at: https://www.mospi.gov.in/global-indicator-frameworkiaeg-sdgs.

Naderi, M. M., Mirchi, A., Bavani, A. R. M., Goharian, E., Madani, K.(2021). System dynamics simulation of regional water supply and demand using a food-energy- water nexus approach: application to Qazvin Plain, Iran. *J. Environ. Manag*. 280, 111843.

Obiasi, A. (2017). Overpopulation: A threat to sustainable agriculture and food security in developing countries? A review. Available at: https://www.researchgate.net/publication/324792051_OVERPOPULATION_A_THREAT_TO_SUSTAINABLE_AGRICULTURE_AND_FOOD_SECURITY_IN_DEVELOPING_COUNTRIES_A_Review.

Ozturk, I. (2015). Sustainability in the food-energy-water nexus: evidence from BRICS (Brazil, the Russian Federation, India, China, and South Africa) countries. *Energy* 93, 999–1010.

Paroda, R. S. and Joshi. P. K. (eds) (2017). Proceedings of the National Conference on Sustainable Development Goals: India's Preparedness and the Role of Agriculture.

New Delhi, India, May 11–12, 2017. Indian Council of Agricultural Research (ICAR), Trust for Advancement of Agricultural Sciences (TAAS) and International Food Policy Research Institute (IFPRI). 48 pp.

Paroda, R. S. (2018). Reorienting Indian Agriculture: Challenges and Opportunities. *Food Sec*. 11:985-986. https://doi.org/10.1007/s12571-019-00955-6.

Paoli, A. D., Addeo, F. (2019). Assessing SDGs: A methodology to measure sustainability. *Athens Journal of Social Science*. 6(3):229-250. https://doi.org/10.30958/ajss.6-3-4.

Poore, J., Nemecek, T. (2018) Reducing food's environmental impacts through producers and consumers. *Science* 360:987–992. https://doi.org/10.1126/science.aaq0216.

Priyadarshini, P., Abhilash, P. C. (2020). Policy recommendations for enabling transition towards sustainable agriculture in India. *Land Use Policy*, 96: 104718-104732. https://doi.org/10.1016/j.landusepol.2020.104718.

Ramankutty, N., Mehrabi, Z., Waha, K., Jarvis, L., Kremen, C., Herrero, M., and Rieseberg, L. H. (2018). Trends in global agricultural land use: Implications for environmental health and food security. *Annual Review of Plant Biology,* 69(1), 789-815. https://doi.org/10.1146/annurev-arplant-042817-040256.

Rao, S. C., Kareemulla, K., Krishnan, P., Murthy, G. R. K., Ramesh, P., Ananthan, P. S., Joshi, P. K. (2018). Agro-ecosystem based sustainability indicators for climate resilient agriculture in India: a conceptual framework. *Ecol. Indic*. https://doi.org/10.1016/j. ecolind.2018.06.038.

Reinales, D, Zambrana-Vasques, D., Saez-De-Guinoa, A. (2020). Social life cycle assessment of product value chains under a circular economy approach: A case study in plastic packaging sector. *Sustainability*. 12(16), 6671. https://doi.org/10.3390/su12166671.

Rivero, R. M., Mittler, R., Blumwald, E., Zandalinas, S. I. (2021). Developing climate-resilient crops: Improving plant tolerance to stress combination. *Plant Journal.* 109(2):373-389. https://doi.org/10.1111/tpj.15483.

Robinson, S., Mason d'Croz, D., Islam, S., Sulser, T. B., Robertson, R. D., Zhu, T., et al. (2015). The International Model for Policy Analysis of Agricultural Commodities and Trade (IMPACT): Model description for Version 3 (IFPRI Discussion Paper No. 1483). Washington, DC: International Food Policy Research Institute (IFPRI). Retrieved from http://ebrary.ifpri.org/cdm/singleitem/collection/p15738coll2/id/129825/rec/1.

Rockstrom, J., Williams, J., Daily, G., Noble, A., Matthews, N., Gordon, L. (2017). Sustainable intensification of agriculture for human prosperity and global sustainability. *Ambio*, 46(1), 4-17. https://doi.org/10.1007/s13280-016-0793-6.

Roy, A., Pramanick, K. (2019). Analysing progress of sustainable development goal 6 in India: Past and future. *J Environmental Mgmt.ens,* 232, 1049-1065. https://doi.org/10.1016/j.jenvman.2018.11.060.

Sachs, J. D., Lafortune, G., Fuller, G., Drumm, E. (2023) Sustainable Development report 2023 Implementing the SDG stimulus. Available at: https://dashboards.sdgindex.org/.

Schaldach, R., Alcamo, J., Koch, J., Kölking, C., Lapola, D. M., Schüngel, J., & Priess, J. A. (2011). An integrated approach to modelling land-use change on continental and global scales. *Environmental Modelling & Software*, 26(8), 1041–1051. https://doi.org/10.1016/j.envsoft.2011.02.013.

Selvaraj, A., Gautam, J., Verma, S., Verma, G., Jain, S. (2021). Life cycle sustainability assessment of crops in India. Curr. Res. *Envt. Sust.* 1-21. https://doi.org/10.1016/j.crsust.2021.100074.

Springman, M., Clark, M., Mason-D'Croz, D., Wiebe, K., Biodirsky, B. L., Lassaletta, L. (2018). Options for keeping the food system within environmental limits. *Nature,* 562(7728), 519-525. https://doi.org/10.1038/s41586-018-0594-0.

Srivastava, P., Singh, R., Tripathi, S., and Singh, R. A. (2016). An urgent need for sustainable thinking in Agriculture- An Indian scenario. *Ecological indicators*, 67, 611-622. https://doi.org/10.1016/j.ecolind.2016.03.015.

Sternet, T., Barbier, E. B., Bateman, I., et al. (2019). Policy design for the anthropocene. *Nat. Sustain.* 2, 14–21. https://doi.org/10.1038/s41893-018-0194-x.

Tilman, D., Clark, M., Williams, D. R., Kimmel, K., Polasky, S., and Packer, C. (2017). Future threats to biodiversity and pathways to their prevention. *Nature,* 546(7656), 73-81. https://doi.org/10.1038/nature22900.

United Nations (2023). Sustainable Development Goals report 2023: Special edition Towards a rescue plan for people and planet. https://desapublications.un.org/publications/sustainable-development-goals-report-2023-special-edition.

Voglhuber-Slavinsky, A., Zicari, A., Smetana, S., Moller, B., Donitz, E., Vranken, L., Zdravkovic, M., Aganovic, K., Bahrs, E. (2022). Setting life cycle assessment (LCA) in a future-oriented context: the combination of qualitative scenarios and LCA in the agri-food sector. *European Journal of Futures Research,* 10(15) https://doi.org/10.1186/s40309-022-00203-9.

Wada, Y., Gleeson, T., Esnault, L. (2014). Wedge approaches to water stress. Nat. Geosci. 7, 615–617. https://doi.org/10.1038/ngeo2241.

Yue, Q., Wu, H., Wang, Y., Guo, P. (2021). Achieving sustainable development goals in agricultural energy-water-food nexus system: An integrated inexact multi-objective optimization approach. *Resources ConservRecycling.ens.* 174, 105833. https://doi.org/10.1016/j.resconrec.2021.105833.

Zhang, X., Yao, G., Vishwakarma, S., Dalin, C., Komarek, A. M., Kanter, D. R., Davis, K. F., Pfeifer, K., Zhao, J., Zou, T., D'Odorica, P., Folberth, C., Rodriguez, F. G., Fanzo, J., Rosa, L., Dennison, W., Musumba, M., Heyman, A., Davidson, E. A. (2021). Quantitative assessment of agricultural sustainability reveals divergent priorities among nations. *One Earth.* 4: 1262-1277. https://doi.org/10.1016/j.oneear.2021.08.015.

## Chapter 3

# Embracing Technological Disruption: Digital Tools for Extension

**Shivam Singh**[1,*]
**Harish Chandra Singh**[1]
**Pawan Kumar Gupta**[1]
**Aman Verma**[2]
**and Kapil Verma**[2]

[1]Department of Agricultural Extension Education, Chandra Shekhar Azad University of Agriculture and Technology, Kanpur, Uttar Pradesh, India
[2]Department of Extension Education Acharya Narendra Deva University of Agriculture and Technology Kumarganj, Ayodhya, Uttar Pradesh, India

### Abstract

This chapter explores the transformative impact of digital technologies on agricultural extension services. As the world rapidly embraces technological innovation, the farm sector stands at the cusp of a digital revolution. We examine how new digital platforms, mobile applications, and integrated information and communication technologies are reshaping the landscape of agricultural knowledge dissemination and farmer support. Our analysis begins by outlining the current technological disruptions in agriculture and their potential to revolutionize extension services. We then investigate various digital tools, including mobile apps, online platforms, remote sensing technologies, and data analytics. We assess their effectiveness in enhancing outreach to farmers, improving knowledge transfer, and

* Corresponding Author's Email: shivam609patel@gmail.com

In: Global Horizons in Agricultural Extension Education
Editors: R. K. Doharey, Aman Verma, Satyapriya et al.
ISBN: 979-8-89530-446-4

promoting sustainable agricultural practices. Drawing on a range of case studies and practical examples, we highlight the opportunities and challenges associated with incorporating digital tools into traditional extension systems. Key considerations such as accessibility, usability, and infrastructure constraints in rural contexts are discussed in depth. The discussion concludes by synthesizing insights and offering practical recommendations for policymakers, extension practitioners, and other stakeholders. These recommendations aim to guide the effective leveraging of technological innovations to enhance agricultural development and strengthen the resilience of rural communities in this digital era. By providing a comprehensive overview of the digital transformation in agricultural extension, this work serves as a valuable resource for professionals, researchers, and students interested in the intersection of technology and agricultural development.

**Keywords:** technological, sustainable, papulation, geographical boundaries

## 3.1. Introduction

Farmers require timely and accurate information to enhance their agricultural practices, productivity, and sustainability. Traditionally, agricultural extension services have served this purpose through workshops, field visits, and direct interactions with experts. However, the advent of digital technologies has transformed the landscape of extension services, introducing mobile apps, websites, and other tools that provide real-time information. For instance, farmers now use mobile devices to access weather updates, pest management techniques, and soil health insights through specialized sensors. These digital tools offer significant advantages, including broader reach, faster dissemination of information, peer-to-peer learning, and advanced decision-making capabilities. However, challenges such as unequal access, limited digital literacy, privacy concerns, and ensuring equitable benefits remain significant barriers to fully realizing the potential of these technologies.

The past decade has witnessed an unprecedented integration of digital technologies across various sectors, driven by innovations that have reshaped industries and organizational structures (Yoo et al., 2010). Digital transformation, characterized by the adoption of cutting-edge technologies, is no longer confined to high-tech industries but has permeated businesses of all sizes and sectors (Warner & Wäger, 2019). This transformation emphasizes the role of dynamic capabilities, including technological entrepreneurship and market expansion, in creating value at both organizational and societal levels (Bouwman et al., 2018).

In agriculture, extension services in advanced countries have traditionally followed a top-down approach, focusing on disseminating technical advice to increase production and profitability. While effective for wealthier farmers, this model often excludes resource-poor farmers (Chambers & Ghildyal, 1985). Over time, agricultural extension has shifted toward more participatory and experiential learning approaches, emphasizing farmer-to-farmer knowledge exchange and community-led strategies (Francis & Carter, 2001). Despite this shift, practical and logistical challenges persist, limiting the adoption of these approaches at scale.

The integration of agroclimatic information into extension services is essential for supporting farmers' decision-making and risk management amidst challenges like population growth, climate change, and food security concerns (Australian Academy of Sciences, 2010; Haigh et al., 2015). Modern agricultural extension must combine insights from climate science, education, and digital technologies to provide timely, tailored, and cost-effective solutions.

Digital information and communication technologies (ICTs) have significantly enhanced the reach and effectiveness of extension services. Early applications of mobile technology, such as voice and SMS-based m-agriculture tools, paved the way for more advanced solutions with the advent of smartphones and affordable data services (Aker, 2011; Gichamba & Lukandu, 2012). These technologies now enable media-rich, context-specific content delivery and sophisticated decision-support tools employing biophysical modeling to optimize farm management (McCown, 2002).

Despite these technological advancements, ICTs have not produced the anticipated increase in innovation adoption rates on farms. Sulaiman and colleagues (2012) argue that this is partially because ICTs are still primarily utilized for conventional tasks such as top-down information dissemination and training. They often fail to integrate the existing knowledge of farmers into the array of contextual factors associated with farm management decision-making, and they fail to facilitate exchanges between farmers (Francis & Carter, 2001). These challenges extend beyond agriculture, as corporations and institutions have also struggled to recognize the importance of interaction in leveraging social media.

## 3.2. Digital Tools for Agriculture Extension

In recent decades, digital technologies have increasingly become integral to propagating agricultural knowledge, offering significant prospects to complement and extend the reach of face-to-face extension services. The emergence of advanced digital platforms, coupled with widespread access to high-speed internet, presents opportunities to explore novel approaches for information exchange in agricultural extension. However, the perceived limitations of digital technology-based decision support tools in influencing agricultural management decisions among many farmers (Francis & Carter, 2001; McCown et al., 2002) have prompted a reconsideration of the need for tools that better align with farmers' requirements. These tools should encompass diverse information assemblages, reasoning processes, and decision-making styles. Recognizing the critical role of discussion in productive adult learning (Kirkup, 2002), there are calls for enhancing logistical support to better integrate with farmers' natural learning processes through experiential learning and exchanges within informal networks. One approach, piloted in Australian agriculture, involves "kitchen table" discussions, where small groups of farmers engage in face-to-face talks with farming systems specialists to apply decision support model outputs to critical decisions (McCown, 2002). Despite its promise, progress in developing and broadly implementing such productive methods within agricultural extension programs has been limited, underscoring the potential role of digital technologies in addressing these gaps (Stone, 2010).

## 3.3. Meaning and Definition of Embracing Technology

*"Embracing"* means willingly and enthusiastically accepting, supporting, or adopting something. It involves acknowledging and actively incorporating it into one's thoughts, actions, or beliefs with openness and readiness to engage with new ideas, changes, or experiences.

*"Embracing technology"* refers to the enthusiastic acceptance and integration of technological advancements into various aspects of life, such as work, communication, education, and daily activities. It involves adopting new tools, platforms, and digital solutions to enhance efficiency, productivity, and outcomes while maintaining a positive attitude towards innovation and change.

## 3.4. Meaning and Definition of Digital Tools

Digital tools refer to software, applications, platforms, or devices that utilize digital technology to perform tasks, facilitate communication, create content, or streamline processes. These tools leverage the capabilities of computers, the internet, and other digital technologies to enhance productivity, efficiency, and effectiveness across various domains. Digital tools have become invaluable in agricultural extension services, functioning as virtual training platforms that deliver instructional videos and e-learning modules. These resources empower farmers to access the latest agricultural techniques and innovations, enabling them to adopt modern and sustainable farming practices. This uninterrupted capacity-building ensures growers stay informed and equipped to tackle evolving agricultural challenges. Additionally, digital technologies democratize knowledge sharing and training in agriculture. Platforms like webinars, e-learning apps, and mobile tools provide flexible, cost-effective opportunities for farmers and extension agents to acquire new skills and stay updated on the latest advancements. This democratization has accelerated the adoption of improved agricultural techniques and sustainable practices.

Examples of Digital Tools Include:

- *Productivity Software:* Tools like Microsoft Office, Google Workspace, or project management platforms like Trello or Asana help individuals and teams organize tasks, manage projects, and collaborate effectively.
- *Communication Tools*: Email clients, instant messaging apps like Slack or Microsoft Teams, video conferencing software like Zoom or Microsoft Teams, and social media platforms enable communication and collaboration among individuals and groups across different locations.
- *Content creation Tools*: Graphic design software such as Adobe Creative Suite, video editing software like Adobe Premiere Pro or Final Cut Pro, and audio editing tools like Audacity facilitate multimedia content creation.
- *Development Tools:* Integrated development environments (IDEs) like Visual Studio or JetBrains IntelliJ IDEA, version control systems like Git, and debugging tools aid software developers in writing, testing, and debugging code.

- *Data analysis Tools:* Statistical software such as R or Python's Pandas library, data visualization tools like Tableau or Power BI, and database management systems like MySQL or MongoDB help analyze and interpret data.
- *Cybersecurity Tools*: Antivirus software, firewalls, encryption tools, and intrusion detection systems protect digital assets and networks from cyber threats.
- *Educational Tools:* Learning management systems (LMS) like Moodle or Canvas, interactive whiteboards, and educational apps help educators deliver content, assess student progress, and facilitate learning.

Overall, digital tools play a crucial role in modern life, enabling individuals and organizations to work more efficiently, communicate effectively, and accomplish tasks that would be challenging or impossible without digital technology.

### 3.4.1. Productivity Software

Microsoft Office is a versatile suite of productivity applications developed by Microsoft Corporation. Key components include:

- *Word:* A widely used word processing program for creating documents ranging from simple letters to complex reports.
- *Excel:* A powerful spreadsheet application used for data analysis, calculations, and visualizations, featuring advanced tools for financial modeling, data manipulation, and statistical analysis.
- *PowerPoint:* A presentation software designed to create visually engaging slideshows for business, academic, and other professional settings, offering tools to incorporate text, images, charts, and multimedia elements.
- *Outlook:* An email client and personal information manager that combines email communication, calendar scheduling, task tracking, and contact organization.

#### *3.4.1.1. Google Workspace*

Formerly known as G Suite, Google Workspace is a cloud-based productivity suite that promotes seamless collaboration and integration across Google services. Key features include:

- *Google Docs:* A web-based word processor enabling real-time collaboration, where multiple users can simultaneously edit documents.
- *Google Sheets:* A cloud-based spreadsheet application similar to Excel, allowing users to analyze data, create charts, and collaborate in real-time.
- *Google Slides:* A presentation tool for creating, editing, and sharing presentations online, with features for adding text, images, animations, and embedded videos.
- *Gmail:* An email service offering a conversation view, customizable filters, and integrated chat functionality to streamline communication.

#### *3.4.1.2. Project Management Tools*

Project management tools are designed to streamline task management, project tracking, and team collaboration. Popular examples include:

- *Trello:* A card-based system for organizing tasks and projects on customizable boards, allowing teams to visualize workflows and track progress effortlessly.
- *Asana:* A comprehensive platform for managing projects, tasks, and timelines, featuring tools for setting goals, visualizing dependencies, and tracking milestones.
- *Jira:* A widely used tool in software development, offering features for issue tracking, agile project management, and release planning.
- *Basecamp:* A centralized platform for project communication, file sharing, and task management, known for its intuitive interface and simplicity in fostering team collaboration.

### 3.4.2. Communication Tools

Communication tools are indispensable in both personal and professional spheres, facilitating seamless interaction and collaboration. Email clients serve as the cornerstone of digital communication, with Microsoft Outlook, Gmail, and Apple Mail leading the way. These platforms enable users to receive and manage email messages efficiently. Microsoft Outlook offers a comprehensive suite of features, including email organization, calendar integration, task management, and contact synchronization. Gmail, renowned for its user-friendly interface and powerful search capabilities, provides robust email management tools such as labels, filters, and priority into inboxes. Apple Mail, the default email client for Apple devices, offers seamless integration with iCloud and intuitive features like smart mailboxes and VIP contacts.

Instant messaging apps have revolutionized real-time communication and collaboration within teams and communities. Platforms like Slack, Microsoft Teams, Discord, and WhatsApp facilitate instant messaging, file sharing, and collaborative workspaces. Slack is renowned for its organized channels, searchable message archives, and integrations with various third-party tools and services. Microsoft Teams combines chat, video conferencing, file storage, and application integration within the Microsoft 365 ecosystem, fostering seamless collaboration among team members. It has evolved into a versatile platform for text, voice, and video communication, enabling communities to connect and collaborate across various interests. WhatsApp, widely used for personal and business communication, offers encrypted messaging, voice calls, video calls, and file sharing across mobile devices and desktops.

Video conferencing software has become indispensable for remote meetings and virtual events, with Zoom, Microsoft Teams, Google Meet, and Skype leading the way. Zoom, known for its ease of use and scalability, offers features like video meetings, webinars, screen sharing, and recording capabilities, catering to various use cases, from small teams to large-scale conferences. Microsoft Teams integrates video conferencing seamlessly with its collaboration hub, providing high-definition video and audio quality, screen sharing, and live captioning. Integrated with Google Workspace, Google Meet enables users to join meetings directly from their browser or mobile device, with features like scheduling, screen sharing, and live streaming. Skype, a veteran in the video calling space, offers video conferencing, voice calling, and instant messaging across devices, with features like call recording and background blur enhancing the user experience. These communication tools empower individuals and teams to connect, collaborate, and communicate effectively in today's digital landscape.

### 3.4.3. Content Creation Tools

Graphic design software such as Adobe Creative Cloud tools like Photoshop, Illustrator, and InDesign are revered as industry standards for many design tasks. Photoshop offers unparalleled manipulation image editing and manipulation versatility, while Illustrator is preferred for vector graphics creation and digital illustration. On the other hand, InDesign serves as a powerful desktop publishing tool, ideal for creating print and digital media layouts.

In the realm of video editing, Adobe Premiere Pro, Final Cut Pro, and Davinci Resolve stand out as premier choices for professionals and enthusiasts alike. Adobe Premiere Pro boasts an extensive set of editing tools, including seamless consolidation with other Adobe Creative Cloud applications and advanced video editing and post-production features. Final Cut Pro, designed exclusively for macOS, offers a streamlined interface, powerful editing capabilities, and optimization for Apple hardware, making it a favorite among Mac users. Davinci Resolve, available in both free and paid versions, provides professional-grade editing, color correction, visual effects, and audio post-production capabilities in a single application, catering to a wide range of video editing needs.

Tools like Audacity, Adobe Audition, and GarageBand for audio editing tasks offer robust features for recording, editing, and mixing audio tracks. Audacity, an open-source audio editor, provides a plethora of tools for audio manipulation, including effects, noise reduction, and multi-track editing, making it a popular choice for users seeking a free solution. Adobe Audition, part of the Adobe Creative Cloud suite, offers professional-grade audio editing and mixing tools and advanced features for audio restoration, noise reduction, and spectral editing. Developed by Apple, GarageBand offers a user-friendly interface and a wide range of virtual instruments, loops, and effects, making it an excellent choice for music composition, recording, and editing tasks, particularly for macOS and iOS users. These audio editing tools empower users to create, edit, and refine audio content with precision and creativity.

### 3.4.4. Development Tools

Integrated Development Environments (IDEs) play a pivotal role in software development, offering comprehensive environments for programmers to write, debug, and deploy code efficiently. Visual Studio, JetBrains IntelliJ

IDEA, and Eclipse stand out as prominent examples, providing a range of features such as code editing, debugging, and version control integration. Visual Studio, developed by Microsoft, offers a rich set of tools for various programming languages and platforms, catering to both Windows and web development. IntelliJ IDEA, developed by JetBrains, provides advanced code assistance, refactoring and built-in tools for JVM-based languages like Java, Kotlin, and Scala. Eclipse, known for its extensibility and versatility, offers a customizable development environment with support for multiple programming languages and frameworks, making it a popular choice among developers.

Version Control Systems (VCS) are essential for managing changes to source code, coordinating work among multiple developers, and facilitating collaboration in software development projects. Git, SVN (Subversion), and Mercurial are widely used VCS platforms, each offering unique advantages and workflows. Git, developed by Linus Torvalds, is renowned for its distributed nature, speed, and branching capabilities, making it the de facto standard for version control in open-source and enterprise projects. SVN (Subversion) is a centralized VCS known for its simplicity and ease of use, which is ideal for projects with a linear development history and centralized control. Mercurial offers a distributed approach like Git, focusing on simplicity and ease of use, making it a suitable choice for projects of all sizes.

Debugging tools are essential for identifying and fixing errors in software code, ensuring the reliability and performance of applications. Debuggers like GDB (GNU Debugger) and Xcode Debugger provide developers with powerful tools for inspecting variables, stepping through code execution, and analyzing program behavior. GDB, a command-line debugger, is widely used in Unix-like operating systems and supports multiple programming languages, including C, C++, and Fortran. Integrated into Apple's Xcode IDE, Xcode Debugger offers a graphical interface for debugging macOS, iOS, watch OS, and TV OS applications, with features like breakpoints, watchpoints, and real-time debugging. These debugging tools empower developers to diagnose and resolve issues efficiently, improving the quality and stability of software applications.

### 3.4.5. Data Analysis Tools

Statistical software like R, Python with pandas and NumPy, and SPSS are essential for statistical analysis and modeling. Data visualization tools like

Tableau, Power BI, and Google Data Studio enable interactive visualizations. Database Management Systems like MySQL, PostgreSQL, MongoDB, and Microsoft SQL Server provide platforms for storing and managing data.

### 3.4.6. Cybersecurity Tools

Antivirus software like McAfee, Norton, and Bitdefender protect against malware and viruses. Firewalls monitor and control network traffic to prevent unauthorized access. Encryption tools such as VeraCrypt and BitLocker secure data from unauthorized access. Intrusion Detection Systems like Snort and Suricata monitor network traffic for suspicious activity.

### 3.4.7. Educational Tools

- *Learning Management Systems (LMS):* Moodle, Canvas, Blackboard, and Schoology provide platforms for delivering online courses, managing learning materials, and assessing student performance.
- *Interactive Whiteboards:* Tools like SMART Board enhance classroom presentations and facilitate interactive learning.
- *Educational Apps:* Khan Academy, Duolingo, and Quizlet offer interactive learning experiences, practice exercises, and educational content across various subjects and disciplines.

These digital tools serve diverse purposes across different industries and domains, empowering individuals and organizations to accomplish tasks efficiently, communicate effectively, and achieve their goals.

### 3.4.8. "e-Krishak Samadhan" Web Portal in India

The web portal named "e-krishak Samadhan" was launched by the government of Uttar Pradesh, India to disseminate information regarding agriculture, advisory services, and problem-resolution support to farmers. This web portal grants farmers the ability to submit their questions and issues related to agriculture online, which are then conveyed by a team of agricultural experts and extension officers to resolve problems. The farmers can also browse a huge repository of agricultural resources, videos, guidelines, and market-related information. The e-Krishak Samadhan web portal remarkably enhances the attainability of information related to agriculture including advisory services for farmers in the Uttar Pradesh state. Whereas farmers revealed that they received prompt replies to their problems, leading to better decision making and enhanced farm management practices. It also promotes peer learning as farmers can access details shared by others, even though it has contributed to diminishing the dependency on conventional extension services. Generally, the e-Krishak Samadhan portal illustrated how digital technologies can augment agriculture extension services' efficiency and effectiveness, respectively.

## 3.5. List of Embracing Technology Tools of Extension

- *Mobile Applications:* Developing and utilizing mobile apps to provide farmers instant access to information on weather forecasts, crop management techniques, market prices, and best agricultural practices.
- *Online Platforms:* Creating online platforms or websites where farmers can access resources, participate in forums, and receive virtual training sessions on various agricultural topics.
- *Remote Sensing Technologies:* Using remote sensing technologies such as drones or satellite imagery to monitor crop health, soil moisture levels, and environmental conditions, allowing for more accurate decision-making and precision agriculture practices.
- *Data Analytics:* Analysing agricultural data from various sources to identify trends, patterns, and insights that can inform decision-making and optimize agricultural practices.
- *Digital Communication*: Leveraging digital communication channels such as email, social media, and messaging apps to provide real-time advisory services, answer queries, and engage with farmers remotely.
- *Webinars and Online Workshops*: Organizing webinars, workshops and training sessions to disseminate knowledge, share best practices, and provide hands-on learning experiences to farmers and extension workers.
- *Virtual Reality (VR) and Augmented Reality (AR):* Utilizing VR and AR technologies to simulate agricultural scenarios, provide

immersive training experiences, and demonstrate complex concepts more interactively and engagingly.

- *Internet of Things (IoT):* Implementing IoT devices and sensors on farms to collect data on crop growth, soil conditions, and environmental factors, enabling farmers to monitor and manage their operations more efficiently.
- *Blockchain Technology:* Exploring the use of blockchain technology for traceability and transparency in agricultural supply chains, ensuring the authenticity and quality of agricultural products from farm to fork.
- *E-Learning Platforms:* Developing e-learning platforms or online courses tailored to the specific needs and interests of farmers, offering modules on topics such as sustainable farming practices, pest management, and crop rotation techniques.

## 3.6. Selection Criteria for Embracing Technology

Technology selection should focus on those technologies most suited to the learning structure. For example, formal workshop learning aligns well with synchronous Internet sessions such as Zoom or learning management systems such as Blackboard. Informal learning can be facilitated with collaborative tools such as Google Docs and Now Comment Information aggregation tools (e.g., content-specific apps, infographics) or social media platforms (e.g. Twitter), which can facilitate independent learning and connections beyond the organization.

The selection criteria for embracing technology in extension services typically involve several vital factors to ensure the chosen technologies are effective, practical, and beneficial for farmers and extension workers. Here are some standard selection criteria:

- *Relevance to Agricultural Needs*: The technology should address specific challenges and needs faced by farmers and extension services in the agricultural sector. It should provide solutions or tools that directly improve farming practices' productivity, sustainability, and profitability.
- *Ease of Use*: The technology should be user-friendly and intuitive, accessible to farmers with varying levels of digital literacy. Complex

technologies may deter adoption, so simplicity and ease of use are crucial considerations.

- *Accessibility:* The technology should be accessible to farmers, considering factors such as internet connectivity, device compatibility, and affordability. It should be available across different platforms and devices to ensure widespread adoption and usage.
- *Scalability:* The technology should potentially scale up and reach many farmers across different regions or communities. Scalability ensures that the benefits of the technology can be maximized and extended to a broader audience.
- *Compatibility with Existing Infrastructure*: The technology should be compatible with existing infrastructure and systems used by farmers and extension services. It should seamlessly integrate with other tools or platforms, minimizing disruptions and compatibility issues.
- *Cost-effectiveness:* The technology should offer value for money with a reasonable cost-benefit ratio. Considerations should be made regarding the initial investment, ongoing maintenance costs, and potential long-term benefits or savings the technology generates.
- *Data Privacy and Security*: The technology should adhere to strict data privacy and security standards to protect sensitive information shared by farmers and extension workers. Measures should be in place to safeguard data against unauthorized access, breaches, or misuse.
- *Evidence of Effectiveness*: There should be evidence or case studies demonstrating the effectiveness and impact of the technology in similar contexts or settings. Empirical research, user testimonials, and pilot studies can provide valuable insights into adopting the technology's potential benefits and outcomes.
- *Support and Training:* Adequate support, training, and technical assistance should be available to farmers and extension workers to ensure they can effectively use and maximize the benefits of the technology. Training programs and user manuals should be provided to facilitate learning and skill development.
- *Sustainability:* The technology should be long-term sustainable, with ongoing support, updates, and maintenance provided by the technology provider or developer. Considerations should be made regarding the technology's long-term viability in evolving agricultural landscapes.

## 3.7. Challenges and Opportunities

India encounters hurdles in embracing technology because of limited income, illiteracy, and small land holdings, which impede the adoption of modern agricultural practices. Despite the evident advantages of digital technologies, obstacles exist that require attention. Challenges like poor connectivity in rural regions, farmers' lack of digital literacy, and apprehensions regarding data privacy must be addressed. Collaborative initiatives involving governments, private enterprises, and NGOs hold the potential to address these barriers, fostering opportunities to narrow disparities and certifying a fair approach towards digital agricultural extension facilities.

## Conclusion

Digital tools convert agricultural extension facilities and allow farmers easy access to information including timely advisory services and personalized recommendations. However, the manifestation of data analytics has enabled data-driven decision-making and precision agriculture. Moreover, the digital revolution in agricultural extension presents unprecedented opportunities to modernize traditional farming practices. The growers and extension agents can access real-time information through digital tools, improve decision-making, and embrace sustainable agricultural practices. Furthermore, addressing challenges like ease of access and data security is vital for ensuring unbiased digital benefits disbursement in agriculture extension. Digital technologies, including remote sensing, data analytics, and mobile applications, empower growers with real-time insights on climate, pests, soil, market prices, enabling informed decisions and enhanced resource allocation.

Moreover, digital integration expands to outstretch the agricultural extension facilities, fostering knowledge diffusion and innovation among farmers. Finally, this research contributes to sustainable agricultural transformation and ensures food security and rural development. Inclusively, these tools have elevated productivity and promoted sustainability, including revolutionizing farmer engagement with extension services and the agricultural sector.

## References


Aker JC, Ghosh I, Burrell J (2016). The Promise (and Pitfalls) of ICT for Agriculture Initiatives. *Agricultural Economics*, 431.

Aker, J. C. (2011). Dial "A" for agriculture: A review of information and communication technologies for agricultural extension in developing countries. *Agricultural Economics*, 42(6), 631–647.

Allevato N, Kagwanja J, Bonomi G, Porcari A, Andreini M, (2016), The M-Kilimo: Mobile phone-based system to improve smallholder farmers' access to agricultural information. In The Impact of ICT on Quality of Working Life. *Springer, Cham,* 99-116.

Australian Academy of Sciences. (2010). Climate change challenges to health: Risks and opportunities. *Canberra: Australian Academy of Science.*

Babu, S., Glendenning, C. J., Asenso-Okyere, K., & Govindarajan, V. S. (2012). Farmers' information needs and search behaviors: Case study in Tamil Nadu, India. *Journal of Agricultural Education and Extension*, 18(4), 359–376.

Barnwell, P. (2023). Embracing Technologies with Purpose. *Council Chronicle*, *33*(1), 28-33.

Barton, E. T., Barton, E. A., Barton, S., Boyer, C. R., Brosnan, J., Hill, P., ... & Stafne, E. (2017). Using technology to enhance extension education and outreach. *Hort Technology*, *27*(2), 177-186.

Beaudoin, D. (2008). Exploring virtual worlds for informal learning: Ethnographic case studies. *Computers & Education,* 50(1), 398–407.

Bhardwaj A, Pal A. (2020), Digital India initiative: A game changer for rural development in India. *Journal of Rural Studies and Research,* 10(1):116-119.

Boag, A. (2014). Social media and engagement. In The Essentials of Social Media Marketing (pp. 111–126). *Palgrave Macmillan, London.*

Bouwman, H., Nikou, S., Molina-Castillo, F. J., & de Reuver, M. (2018). The impact of digitalization on business models. *Digital Policy, Regulation and Governance*, *20*(2), 105-124.

Brennan, L. E., Mee, B. P., Fisher, B. S., Snell, L. A., & Lee, A. K. (2007). Focusing on farmers: Understanding agricultural extension experiences to build effective communication strategies. *Journal of Extension,* 45(1), 1–11.

Carter, M., & Francis, R. (2001). Mentoring and beginning teachers' workplace learning. *Asia-Pacific Journal of Teacher Education*, *29*(3), 249-262.

Ch Srinivasa Rao PD, Sreekanth and GRK, Murthy Status, (2023), Scope of Policy Interventions in Digital Agriculture in India. National Symposium on Digital Farming: The Future of Indian Agriculture, ICAR-IISS, Bhopal; 2023.

Chambers, R., & Ghildyal, B. P. (1985). Agricultural extension: Worldwide institutional evolution and forces for change. *Agricultural Administration*, 20(3), 131–158.

Chambers, R., & Ghildyal, B. P. (1985). Agricultural research for resource-poor farmers: the farmer-first-and-last model. *Agricultural administration*, *20*(1), 1-30.

Crittenden, W. F., Biel, I. K., & Lovely III, W. A. (2019). Embracing digitalization: Student learning and new technologies. *Journal of marketing education*, *41*(1), 5-14.

Diem, R., Hino, J., Martin, W., & Meisenbach, T. (2011). Mobile phone use by farmers in Burkina Faso. Development Informatics Working Paper Series, 52, 1–21.

Francis, C., & Carter, S. (2001). Farmer and researcher partnerships in sustainable agriculture: Some lessons for research, extension, and education. *American Journal of Alternative Agriculture,* 16(3), 125–129.

Gebbers R, Adamchuk VI, (2010), Precision agriculture and food security. Science, 327(5967):828-831.

Gichamba, A., & Lukandu, I. A. (2012). A model for designing M-agriculture applications for dairy farming. *The African Journal of Information Systems*, *4*(4), 1.

Gichamba, D. M., & Lukandu, R. O. (2012). Towards an m-agriculture framework for developing countries: A Kenyan case. *International Journal of Information Communication Technologies and Human Development (IJICTHD)*, 4(1), 68–82.

Graeub, B. E., Chappell, M. J., Wittman, H., Ledermann, S., Kerr, R. B., & Gemmill-Herren, B. (2015). The state of family farms in the world. *World Development,* 87, 1–15.

Khalil A, Lei S, Arshad K, The Role of Elearning in agriculture: Prospects and challenges. *International Journal of Agriculture and Biology.* 2019;21(4): 809-818.

Kirkup, L. (2002). Adult learning in the digital age: Information technology and the learning society. *Routledge.*

Kumar, R. N., Rao, C. R., Raju, B. M. K., Nirmala, G., Pankaj, P. K., & Sailaja, B. (2022). impact assessment of use of digital tools in extension: case studies-review. *Tools on Monitoring, Evaluation and Impact Assessment of Rainfed Agricultural Programs*, 62.

Lowder, S. K., Skoet, J., & Singh, S. (2014). What do we really know about the number and distribution of farms and family farms in the world? Background paper for The State of Food and Agriculture 2014. ESA Working Paper No. 14-02. *FAO, Rome.*

McCown, R. L. (2002). Changing systems for supporting farmers' decisions: problems, paradigms, and prospects. *Agricultural systems*, *74*(1), 179-220.

McCown, R. L. (2002). How do we deliver together the holy grail of science and profit to farmers? In *Proceedings of the 10th Australian Agronomy Conference*, Hobart, Tasmania, Australia.

Mee, B. P., Katz, R. W., Alem, L., & Kravis, J. (2007). A review of agricultural information dissemination systems in Australia: Towards improving the knowledge base of practitioners. *Australian Journal of Agricultural Research*, 58(8), 737–754.

Mee, B. P., Turpin, T., Upton, J., Holden, N. M., & Hennessey, D. (2007). Collaborative learning: Farmer-to-farmer extension in Ireland. Journal of Extension, 45(6), 1–9.

Nelson, R., Holzworth, L., Hammer, G., & Hayman, P. (2002). Infusing learning theory into learning networks. Agricultural Systems, 73(1), 271–295.

Robbin, A. (2011). Embracing technology and the challenges of complexity. *tripleC: Communication, Capitalism & Critique. Open Access Journal for a Global Sustainable Information Society*, *9*(1), 11-27.

Singh A, Singh R, Singh G, (2020), An Impact evaluation of e-krishak samadhan web portal on farmers in Uttar Pradesh, India. *Agricultural Economics Research Review*, 33(Conferenc), 157-164.

Singh, N. K., Sunitha, N. H., Tripathi, G., Saikanth, D. R. K., Sharma, A., Jose, A. E., & Mary, M. V. (2023). Impact of Digital Technologies in Agricultural Extension. *Asian Journal of Agricultural Extension, Economics & Sociology*, *41*(9), 963-970.

Stone, A. M. (2010). E-learning, agricultural extension, and family farming in rural Peru. *Journal of Extension,* 48(5), 1–5.

Warner, K. D. (2006). A learner-centered approach to agriculture education: Integrating practices and adult learning theories. *Journal of Agricultural Education,* 47(3), 87–97.

Warner, Karl SR, and Maximilian Wäger. (2019) "Building dynamic capabilities for digital transformation: An ongoing process of strategic renewal." *Long range planning,* 326-349.

Wiesböck, F., & Hess, T. (2020). Digital innovations: Embedding in organizations. *Electronic Markets*, *30*(1), 75-86.

Yadav R, Ranjan R, Singh (2021), A, Role of ICT in agriculture extension services in India. *Journal of Community Mobilization and Sustainable Development*, 16(1): 86-89.

Yoo, Y., Henfridsson, O., & Lyytinen, K. (2010). Research commentary—the new organizing logic of digital innovation: an agenda for information systems research. *Information systems research*, *21*(4), 724-735.

Yunus, M. M., & Suliman, A. (2014). Information & communication technology (ICT) tools in teaching and learning literature component in Malaysian secondary schools. *Asian social science*, *10*(7), 136.

## Chapter 4

# Building Bridges: Intercultural Communication and Collaboration in Agricultural Extension

**Amrit Warshini[1,*]**
**Vineeta Kumari[2]**
**Sugandha Munshi[3]**
**Vineeta Chandra[4]**
**Aman Verma[1]**
**and Anurag Dixit[1]**

[1]Department of Extension, Acharya Narendra Deva University of Agriculture and Technology, Ayodhya, India
[2]National Institute of Agricultural Extension Management, Hyderabad, India
[3]Internatinal Rice Research Institute, Varanasi, India
[4]Department of Agricultural Extension Education,
Centurion University of Technology and Management, Paralakhemundi, Odisha, India

## Abstract

Practical extension work thrives on reaching diverse audiences and fostering collaboration across cultures. This chapter explores the critical intersection of intercultural communication and cooperation in extension programs. We delve into the importance of cultural competency for extension professionals, equipping them to navigate the nuances of communication styles, values, and beliefs across different communities. The chapter highlights strategies to bridge cultural divides, emphasizing

* Corresponding Author's Email: amritwarshini1312@gmail.com

In: Global Horizons in Agricultural Extension Education
Editors: R. K. Doharey, Aman Verma, Satyapriya et al.
ISBN: 979-8-89530-446-4

active listening, fostering trust, and utilizing culturally appropriate communication methods. We explore how intercultural awareness can enhance collaboration, promoting shared decision-making and fostering a sense of ownership among stakeholders. This chapter provides valuable insights for extension professionals working in increasingly globalized environments, empowering them to deliver impactful programs that resonate with diverse audiences.

**Keywords:** intercultural communication, extension education, collaboration, cultural competency, diversity, culturally appropriate communication

## 4.1. Introduction

Intercultural communication is the exchange of information between individuals or groups from different cultural backgrounds. It is crucial in extension work, where professionals engage with diverse communities to provide education, resources, and support. Cultural differences significantly impact communication and collaboration in these settings, influencing how messages are conveyed, interpreted, and acted upon. Cultural differences can lead to misunderstandings, misinterpretations, and ineffective communication in diverse settings. For example, varying communication styles, non-verbal cues, and language barriers can hinder the exchange of information between extension professionals and community members. These differences can affect extension programs' trust, engagement, and outcomes. Intercultural communication is a multifaceted field that has become increasingly relevant in our interconnected world. It refers to exchanging information and ideas among individuals from different cultural backgrounds, each bringing their unique values, beliefs, norms, and communication styles to the interaction.

In the context of extension work, which aims to disseminate knowledge, technology, and practices to diverse communities, effective intercultural communication is not just important—it is essential. Understanding the significance of intercultural communication in extension work requires exploring how cultural differences impact communication and collaboration. These differences can manifest in various ways, from language barriers to differing attitudes towards authority, time, and decision-making processes. Extension professionals must navigate these cultural nuances to ensure the success and relevance of their programs. Understanding cultural nuances is essential for effective intercultural communication in extension work.

Extension professionals can adapt communication strategies by recognizing and respecting cultural differences to connect with diverse audiences better. This adaptation may involve using appropriate language, considering cultural norms, and being mindful of non-verbal cues when interacting with individuals from different cultural backgrounds. Moreover, cultural differences can influence collaboration in extension work. To foster effective collaboration, navigating cultural norms, values, and communication preferences is essential when working with diverse groups. Extension professionals can build stronger relationships, enhance teamwork, and achieve shared goals with community partners by acknowledging and valuing cultural diversity.

Intercultural communication is a crucial aspect of extension work, enabling professionals to engage with diverse communities and deliver impactful programs. By understanding and addressing cultural differences, extension professionals can enhance communication, collaboration, and, ultimately, the success of their extension initiatives.

In today's globalized world, effective intercultural communication and collaboration are paramount for success in various fields, including extension services. Extension services are crucial in disseminating knowledge, technologies, and practices to diverse communities. However, the effectiveness of these services heavily relies on the ability to navigate cultural differences and foster collaboration across cultures.

## 4.2. Understanding Intercultural Communication in Extension

Intercultural communication within extension services involves interactions among individuals from different cultural backgrounds, each bringing unique perspectives, values, and communication styles. This diversity enriches the field but also presents challenges that must be addressed. Misunderstandings, misinterpretations, and cultural barriers can hinder effective communication and collaboration, ultimately impacting the success of extension programs.

### 4.2.1. Hofstede's Cultural Dimensions

Hofstede's cultural dimensions theory offers valuable insights to comprehend the complexities of intercultural communication. This theory identifies key dimensions of culture, including power distance, individualism-collectivism, masculinity-femininity, uncertainty avoidance, and long-term orientation. Understanding these dimensions helps extension professionals navigate cultural differences and tailor communication strategies accordingly.

For instance, in a high-power distance culture where hierarchical structures are valued, extension agents may need to adjust their communication style to respect authority and establish trust. Conversely, in low power distance cultures, a more egalitarian approach to communication may be practical.

1. *Power Distance Index (PDI)*
   This dimension measures how much a society values hierarchical order and respects authority. A high PDI score indicates that people accept a hierarchical order where everybody has a place, and there is a clear division between superiors and subordinates. A low PDI score suggests a flatter hierarchy and more equality among individuals.
2. *Individualism vs. Collectivism (IDV)*
   This dimension measures whether a society prioritizes individual goals and achievements (individualism) or group goals and collective well-being (collectivism). In individualistic cultures, people tend to look after themselves and their immediate families, while in collectivistic societies, there is a stronger emphasis on group harmony and loyalty.
3. *Masculinity vs. Femininity (MAS)*
   This dimension looks at the distribution of roles between genders. Masculine societies value competitiveness, assertiveness, and material success, while feminine societies value cooperation, modesty, and quality of life.
4. *Uncertainty Avoidance Index (UAI)*
   This dimension measures a society's tolerance for uncertainty and ambiguity. A high UAI score suggests a society that feels threatened by ambiguity and has created rules and regulations to minimize uncertainty. A low UAI score indicates a more relaxed attitude toward ambiguity and greater risk acceptance.
5. *Long-Term Orientation vs. Short-Term Normative Orientation (LTO)*
   This dimension focuses on how much a society values long-standing traditions and norms versus short-term thinking and values. Long-term-oriented societies prioritize persistence, perseverance, and thrift,

while short-term-oriented societies emphasize immediate rewards, respect for tradition, and social obligations.

6. *Indulgence vs. Restraint (IVR)*
   This dimension looks at the extent to which a society allows for the gratification of basic human desires related to enjoying life and having fun. Indulgent societies allow relatively free gratification of natural human desires, whereas restrained societies suppress gratification of needs and regulate it using strict social norms.

### *4.2.1.1. Applications*

#### *4.2.1.1.1. Cultural Difference Awareness*

Geert Hofstede shed light on how cultural differences are still significant today in a world that is becoming more and more diverse.

Hofstede's cultural dimensions can help explain why certain behaviors are common in different cultures.

For example, individualism vs. collectivism can help explain why some cultures place more emphasis on personal achievement than others. Masculinity vs. feminism could help explain why some cultures are more competitive than others. Moreover, long-term vs. short-term orientation can help explain why some cultures focus more on the future than the present (Hofstede, 2011).

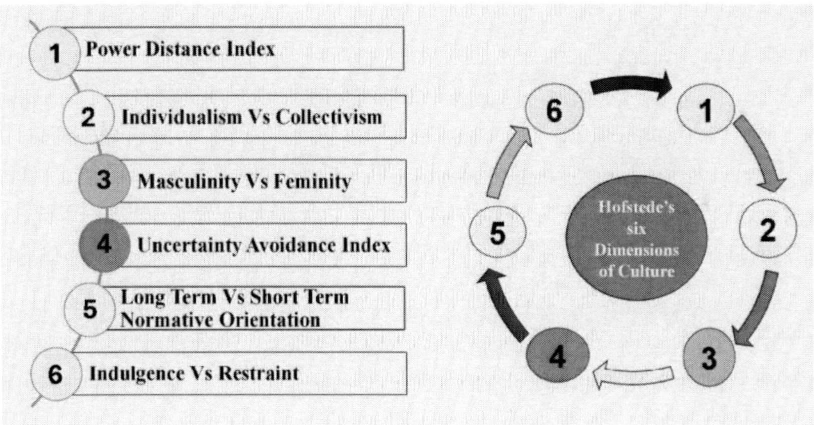


**Figure 4.1.** Shows Hofsted's Cultural Dimensions Theory.

*4.2.1.1.2. International Communication and Negotiation*

Hofstede's cultural dimensions can also predict how people from different cultures interact. For example, if two people from cultures with high levels of power distance meet, they may have difficulty communicating because they have different expectations about who should be in charge (Hofstede, 2011).

### 4.2.2. The Iceberg Theory of Culture

The Iceberg Theory of Culture, also known as Hall's Cultural Iceberg, is a model developed by anthropologist Edward T. Hall to illustrate how culture can be understood through visible and invisible elements. The model uses the metaphor of an iceberg, where only a small portion is visible above the waterline, while most of the iceberg lies beneath the surface.

Here is an explanation of the Iceberg Theory of Culture

- *Visible Culture (Above the Waterline):* This represents the visible aspects of culture, which are easily observable and can include Behaviors, Dress, Food, Language, Artifacts, and Customs. These elements are often the first things we notice when encountering a new culture.
- *Invisible Culture (Below the Waterline):* This refers to the deeper, less visible aspects of culture that are harder to notice but profoundly influence behavior. It includes Beliefs, Values, Assumptions, Norms, and Worldviews.

  These elements are often unconscious and shape how individuals within a culture perceive the world and interact with others.
- *Implications:* Understanding the Iceberg Theory of Culture is crucial for effective intercultural communication and interaction. Superficial observation of visible culture may lead to misunderstandings, as the more profound, invisible aspects greatly influence behavior. Successful cross-cultural interactions involve delving beneath the surface to understand the values, beliefs, and assumptions that underpin a culture.

This theory provides a valuable framework for exploring and understanding the complexities of different cultures, emphasizing the importance of looking beyond surface-level observations.

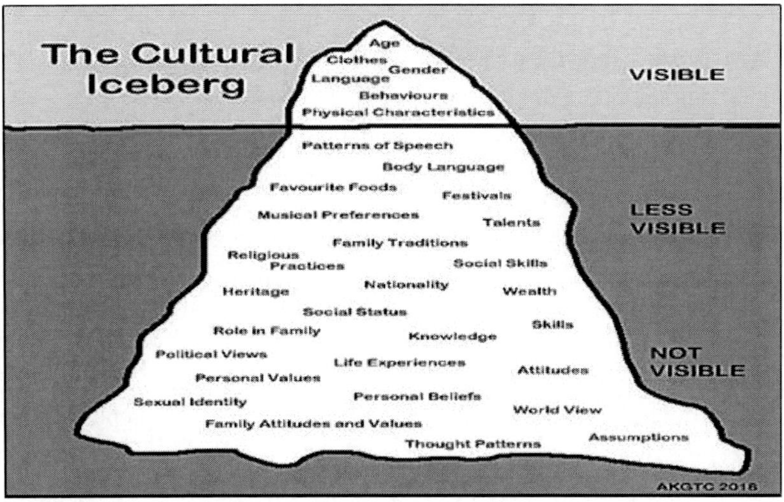


**Figure 4.2.** Shows a model called the Iceberg Theory of Culture.

## 4.2.3. Trompenaars' Cultural Dimensions

Trompenaars' Cultural Dimensions is a framework developed by Fons Trompenaars and Charles Hampden-Turner that helps to understand cultural differences in workplace and societal settings. It identifies seven dimensions along which cultures can be compared, providing insights into how different cultures approach values, communication, and collaboration (Trompenaars, 1998).

### *4.2.3.1. Trompenaars' Seven Cultural Dimensions*

1. *Universalism vs. Particularism*
   - Universalism: Emphasizes rules, laws, and codes of conduct that apply to everyone equally.
   - Particularism: Focuses on relationships and context, where exceptions can be made based on circumstances and relationships.
2. *Individualism vs. Communitarianism*
   - Individualism: Values individual rights, freedom, and personal achievements.
   - Communitarianism: Prioritizes the group or community's interests over individual desires.

3. *Specificity vs. Diffuseness*
   - Specificity: Cultures with a specific orientation are focused and clear in their communication and expectations.
   - Diffuseness: In diffuse cultures, boundaries between work and personal life are less defined, and communication is more indirect.
4. *Neutral vs. Emotional*
   - Neutral: Cultures that value neutrality tends to control emotions, avoid conflict, and maintain a calm demeanor.
   - Emotional: Emphasizes expression of emotions, passion, and open communication of feelings.
5. *Achievement vs. Ascription*
   - Achievement: Based on what an individual has accomplished or their merit.
   - Ascription: Status is given based on age, gender, or position within a hierarchy.
6. *Sequential vs. Synchronic Time*
   - Sequential: Focuses on punctuality, deadlines, and linear progression of tasks.
   - Synchronic: Emphasizes relationships and events simultaneously, valuing flexibility over strict schedule adherence.
7. *Internal vs. External Control*
   - Internal Control: People believe they have control over their fate and outcomes.
   - External Control: Attributes outcomes to external factors such as fate, luck, or the influence of others.

#### *4.2.3.2. Application in Intercultural Communication*

Understanding Trompenaars' Cultural Dimensions can enhance intercultural communication and collaboration by

- *Awareness:* Recognizing cultural differences and adapting communication styles accordingly.
- *Conflict Resolution*: Understanding differing perspectives can lead to more effective conflict resolution.
- *Team Building:* Leveraging cultural diversity for innovation and problem-solving.

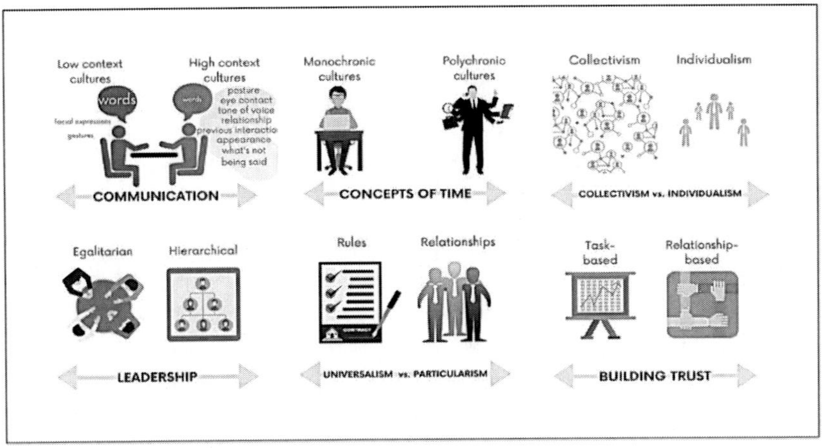


**Figure 4.3.** Shows Intercultural Communication.

## 4.3. Impact of Cultural Differences on Communication

Intercultural communication and collaboration play pivotal roles in the success of extension work, enabling effective engagement with diverse communities. Nevertheless, numerous challenges can impede these efforts.

Language barriers**,** for instance, are a significant hurdle that can obstruct the transmission of ideas and understanding between extension workers and the communities they serve (Gudykunst & Kim, 2017; Watzlawick et al., 2011). A study by Gudykunst and Kim (2017) highlighted how language differences can lead to misinterpretations of technical terms and concepts, thereby hindering the successful delivery of extension services. Overcoming such challenges necessitates a multifaceted approach, including using interpreters, translated materials, and language training for extension professionals.

In addition to language barriers, differences in non-verbal communication present another formidable obstacle in intercultural communication within extension work (Matsumoto & Hwang, 2019; Hall, 1976). Gestures, facial expressions, and body language can vary significantly across cultures, potentially leading to misunderstandings or misinterpretations. For example, an innocuous gesture in one culture may carry a different meaning or even be considered offensive in another (Matsumoto & Hwang, 2019). To navigate these differences effectively, extension workers must develop cultural competence and an understanding of the nuances of non-verbal communication in their communities.

Moreover, misunderstandings arising from differing cultural norms can pose significant challenges (Ting-Toomey & Chung, 2012; Collier, 2016). Cultural norms dictate community behaviors, beliefs, and social practices, influencing how individuals perceive and respond to information. For instance, in some cultures, direct communication may be valued, while in others, indirect communication is preferred to maintain harmony (Ting-Toomey & Chung, 2012). Such differences can lead to misunderstandings or conflicts if not recognized and addressed by extension professionals.

To address these challenges effectively, extension workers must engage in ongoing cultural learning and sensitivity training (Giles & Ogay, 2019; Hammer et al., 2003). This involves developing an awareness of one's own cultural biases and assumptions and actively seeking to understand the cultural norms and communication styles of the communities they work with. Establishing trust and rapport is also essential, creating a foundation for effective communication and collaboration (Collier, 2016). Extension workers can enhance the impact of their programs and services by fostering a culturally inclusive approach, promoting meaningful engagement and sustainable development within diverse communities.

Intercultural communication and collaboration are integral to building relationships and delivering effective services in extension work. However, challenges such as language barriers, non-verbal communication differences, and misunderstandings due to cultural norms must be addressed proactively. Extension professionals can navigate these challenges and create meaningful connections with the communities they serve by investing in cultural competence, training, and building trust.

### 4.3.1. Cultural Sensitivity Training

Cultural sensitivity training has been widely recognized as a valuable tool for improving intercultural communication. Studies have shown its effectiveness in enhancing cultural competence among professionals working in diverse environments (Anderson, 2015). This training helps individuals develop the awareness and skills necessary to interact respectfully and effectively with individuals from different cultural backgrounds (Cushner & Brislin, 1996).

### 4.3.2. Active Listening Techniques

Active listening is a fundamental skill for successful intercultural communication. Research suggests that active listening leads to better understanding and reduces misunderstandings in cross-cultural interactions (Purkey & Schmidt, 1996). By employing techniques such as reflective listening and empathetic responses, extension workers can create a conducive environment for meaningful dialogue (LaFasto & Larson, 2001).

### 4.3.3. Use of Interpreters

Utilizing professional interpreters is crucial for accurate communication in multilingual settings. Studies emphasize the importance of competent interpreters in ensuring that information is conveyed accurately and culturally appropriately (Angelelli, 2004). Interpreters translate language and navigate cultural nuances, ensuring effective communication between parties (Inghilleri, 2003).

### 4.3.4. Diverse Team Composition

Research supports that diverse teams improve outcomes in various contexts, including extension work. Diverse teams are more likely to generate innovative solutions and adapt to complex challenges (Hong & Page, 2004). By bringing together individuals from different cultural backgrounds, extension programs can benefit from various perspectives and experiences (Cox & Blake, 1991).

By integrating these strategies and drawing upon the insights from these diverse studies, extension professionals can enhance their intercultural communication skills and foster collaboration in diverse communities, ultimately leading to more impactful and culturally sensitive extension work.

Intercultural communication and collaboration in extension programs are vital for fostering understanding and achieving positive outcomes in diverse communities. With its rich cultural tapestry and varied contexts, India provides an excellent backdrop for exploring successful instances of such initiatives. This chapter presents compelling case studies and examples highlighting best practices and real-life successes in intercultural communication and collaboration within Indian extension programs.

## 4.4. Tools and Resources for Enhancing Intercultural Communication

Regarding enhancing intercultural communication in extension work, several tools and resources can be invaluable. Here are some examples:

1. *Cultural Intelligence Assessments:* These assessments help extension professionals to understand their own cultural biases and strengths, as well as develop a greater
2. *Cultural Competency Training*: This training equips extension professionals with the knowledge and skills to engage with diverse audiences effectively. They learn about different cultural norms, values, communication styles, and customs, enabling them to adapt their approach accordingly.
3. *Intercultural Communication Workshops*: These workshops provide extension professionals with hands-on practice and guidance in navigating intercultural interactions. They may include activities and simulations that encourage participants to reflect on their cultural assumptions, communicate across cultural differences, and develop strategies for building trust and understanding with diverse individuals and communities.
4. *Cross-Cultural Collaboration Platforms*: These online platforms facilitate communication and collaboration between extension professionals from different cultural backgrounds. By providing a virtual space for sharing ideas, knowledge, and experiences, these platforms enable extension professionals to learn from one another and work together on projects that promote intercultural understanding and collaboration.
5. *Intercultural Communication Guides and Manuals:* These resources provide practical tips, strategies, and best practices for effective intercultural communication in various contexts. They offer insights into cultural differences, communication styles, and etiquette, helping extension professionals navigate intercultural interactions successfully.
6. *Language and Cultural Understanding Programs*: These programs provide extension professionals with language skills and cultural knowledge specific to the populations they serve. By learning the language of the target community and gaining a deeper understanding

of their cultural norms, traditions, and values, extension professionals can communicate more effectively and build stronger relationships with the community members.4. Intercultural Mentoring and Coaching: This approach pairs extension professionals with mentors or coaches who have experience working in diverse cultural contexts. The mentors or coaches provide guidance and support in navigating intercultural interactions, offering insights, strategies, and personal experiences. Intercultural Experiential Learning Activities: These activities provide extension professionals with opportunities to engage in immersive and hands-on experiences that increase their understanding and appreciation of different cultures. These activities may include cultural immersion programs, community service projects in diverse communities, or participation in cultural festivals and events.

## 4.5. Case Study: Intercultural Extension in Action

### 4.5.1. Context

The extension project aimed to improve water management practices in a rural community in India. The team consisted of local farmers, government officials, and international experts. (Global Leadership Foundation 2023).

### 4.5.2. Challenges

- *Language Barriers*: The team members spoke different languages, hindering effective communication.
- *Cultural Norms*: The concept of water conservation varies across cultures.
- *Power Dynamics*: Local farmers initially hesitated to share their traditional knowledge with experts.

#### 4.5.3. Strategies

- *Language Mediation*: Bilingual facilitators facilitated discussions, ensuring everyone understood the content.
- *Community Dialogues*: The team organized community meetings, allowing farmers to express their views and concerns.
- *Knowledge Exchange*: Experts shared scientific insights, while farmers shared indigenous practices.

#### 4.5.4. Expected Outcomes

- *Innovative Solutions*: The collaboration led to hybrid water management techniques that combined scientific principles with local wisdom.
- *Community Ownership*: Farmers actively implemented and maintained the water systems.
- *Policy Influence*: The project findings influenced regional water policies.

Intercultural communication and collaboration are not mere buzzwords but essential competencies for extension professionals. By embracing diversity, fostering dialogue, and adapting to cultural contexts, we can create meaningful impact in extension programs worldwide.

Collaboration "in extension" refers to working together among various stakeholders involved in agricultural extension activities. Agricultural extension involves disseminating agricultural knowledge, technology, and practices to farmers and rural communities. Collaboration in this context involves partnerships and cooperation among entities such as government agencies, NGOs, research institutions, farmers' organizations, and community groups.

### 4.6. Building Intercultural Competence in Extension Professionals

Extension professionals need to develop intercultural competence to effectively engage in intercultural communication and collaboration.

Intercultural competence refers to the ability to effectively navigate and interact with individuals from different cultural backgrounds. It involves having an awareness and understanding of one's cultural values, beliefs, and biases, as well as the ability to adapt and adjust communication styles to accommodate cultural differences. Building intercultural competence starts with self-reflection and awareness. Extension professionals should self-reflect to become aware of their cultural values and biases. They should examine their beliefs, assumptions, and stereotypes about other cultures and be willing to challenge and expand their perspectives. In addition to self-reflection, extension professionals can enhance their intercultural competence through education and training. Intercultural communication and collaboration in extension programs require a deep understanding of cultural differences, effective communication strategies, and the ability to navigate potential barriers and challenges that may arise. To begin building intercultural competence, extension professionals should engage in self-reflection and become aware of their cultural values, biases, and assumptions. They should also seek opportunities for education and training in intercultural communication and cultural competency. Extension professionals are better equipped to communicate and collaborate effectively with diverse audiences by developing intercultural competence.

## 4.7. Key Aspects of Collaboration in Extension

- *Partnership Building*
  We are establishing partnerships and networks among diverse stakeholders to achieve common goals.
  This includes collaboration between government agencies, research institutions, extension workers, farmers' groups, NGOs, and private sector entities.
- *Information Exchange*
  Knowledge, information, and resources are shared among partners. This includes exchanging best practices, research findings, market information, and technological innovations.
- *Joint Decision-Making*
  Involving all stakeholders in the decision-making process related to agricultural extension activities. This ensures that the needs and

priorities of farmers and communities are considered in the planning and implementation of extension programs.

- *Resource Sharing*
  Pooling of resources such as funds, infrastructure, equipment, and expertise. Collaboration allows partners to leverage each other's strengths and resources for more effective and efficient implementation of extension activities.
- *Capacity Building*
  Collaborative efforts to build the capacity of extension workers, farmers, and community members. This includes training programs, workshops, and skill development initiatives to enhance knowledge and skills related to agriculture and rural development.
- *Collective Action*
  Encouraging collective action among farmers and communities for mutual benefit.
  Collaboration fosters the formation of farmer groups, cooperatives, and associations, enabling farmers to work together towards common goals such as accessing markets, bargaining power, and resource sharing.

*Example:* Agricultural Extension Project: A project aimed at promoting sustainable farming practices in a rural community may involve collaboration between government extension agencies, a local NGO specializing in sustainable agriculture, researchers from an agricultural university, and farmers' groups.

1. The government agency provides funding and policy support.
2. The NGO brings expertise in sustainable agriculture techniques.
3. Researchers contribute scientific knowledge and innovations.
4. Farmers' groups provide on-the-ground insights and help disseminate information within the community.

The project can achieve a more significant impact through this collaboration, adopt sustainable practices, and improve farmers' livelihoods.

### 4.7.1. Enhanced Community Engagement

Collaboration, in extension, facilitates enhanced community engagement, which is crucial for the success of programs. When extension professionals work alongside community members, they gain a deeper understanding of local needs, priorities, and challenges. This understanding allows for developing tailored solutions that are culturally sensitive and contextually appropriate. Research by Doe et al. (2019) demonstrates that collaborative extension programs have higher community participation and ownership levels. In a study conducted in rural India, a collaborative extension project involving local farmers, extension agents, and NGOs resulted in increased adoption of sustainable farming practices. This was attributed to the active involvement of community members in decision-making and project implementation.

### 4.7.2. Capacity Building and Knowledge Sharing

Collaboration, in extension, also serves as a platform for capacity building and knowledge sharing. Extension professionals gain access to the latest research findings, innovations, and best practices through partnerships with academic institutions and research organizations. This knowledge is then shared with communities, empowering them to make informed decisions and adopt new technologies. A study by Smith and Brown (2018) highlights the role of collaboration in capacity-building among extension professionals. By engaging in collaborative networks, extension agents have professional development, training, and mentorship opportunities. This not only enhances their skills but also improves the quality of services delivered to communities.

### 4.7.3. Resource Mobilization and Cost Efficiency

Collaboration enables resource mobilization and cost efficiency in extension programs. Partnerships with government agencies and NGOs provide access to funding, equipment, and infrastructure that may not be available otherwise. Pooling resources allows for larger-scale projects and the implementation of comprehensive initiatives.

### 4.7.4. Improved Program Sustainability

Sustainability is a crucial consideration in extension work, and collaboration plays a critical role in ensuring the long-term impact of programs. Sustainable development goals can be better achieved through partnerships that promote environmental stewardship, economic viability, and social equity.

The sustainability outcomes of collaborative extension initiatives in Latin America. Projects that involved partnerships between extension services, local governments, and community organizations showed higher levels of sustainability. These partnerships fostered ongoing support, monitoring, and evaluation, leading to lasting changes in agricultural practices and livelihoods.

The scheme supports capacity building and entrepreneurial activities, encouraging intercultural exchange and collaboration.

1. National Mission on Agricultural Extension and Technology (NMAET)
   - *Objective*: The National Mission on Agricultural Extension and Technology aims to restructure and strengthen agricultural extension to deliver appropriate technology and improved agronomic practices.
   - *Benefits*: The scheme supports the training of farmers and extension agents, which can include modules on intercultural communication and collaboration. Facilitating farmer field schools and demonstrations can bring farmers from diverse backgrounds together to learn and share practices.
2. National Agricultural Market (eNAM)
   - *Objective*: The eNAM scheme aims to create a unified national market for agricultural commodities.
   - *Benefits:* The platform facilitates communication and collaboration between farmers, traders, and buyers from different regions and cultural backgrounds. It encourages transparency and fair pricing, fostering trust among stakeholders.
3. Sub-Mission on Agricultural Extension (SMAE)
   - *Objective*: The SMAE focuses on providing extension services to farmers to increase agricultural production and productivity.
   - *Benefits:* The scheme supports the training of extension workers, which can include components of intercultural communication. It facilitates the dissemination of information and knowledge on agricultural practices among diverse farming communities.

4. National Rural Livelihoods Mission (NRLM)
   - *Objective:* NRLM aims to improve the livelihoods of rural households, including those engaged in agriculture.
   - *Benefits:* NRLM promotes the formation of self-help groups (SHGs) and farmer-producer organizations (FPOs), which can foster collaboration and collective learning among farmers from diverse cultural backgrounds.

While these schemes are not specifically tailored for intercultural communication and collaboration, they provide a framework and resources to support initiatives that promote understanding and collaboration among diverse farming communities. By leveraging these schemes effectively, agricultural extension programs can bridge cultural gaps, facilitate knowledge exchange, and foster sustainable agricultural practices across India. Designing.

## 4.8. Extension Programs for Diverse Audiences

Intercultural communication is vital in extension programs that aim to serve diverse audiences. It is essential to design extension programs that consider the target audience's cultural backgrounds, values, and communication styles. One significant difference between extension programs and traditional university teaching is the audience's involvement in the programs' design and delivery. Extension programs prioritize identifying community needs and then partner with local organizations and individuals to address them. This collaborative approach ensures that the programs are tailored to the specific cultural context of the audience, making them more relevant and effective. By involving the target audience in the process, extension programs can better understand their cultural preferences, communication styles, and unique challenges. This understanding allows extension professionals to adapt their communication strategies, materials, and delivery methods to better resonate with the diverse audience. Intercultural communication in extension programs requires sensitivity and understanding of cultural differences and the ability to navigate potential barriers and challenges.

## 4.9. Evaluating the Impact of Intercultural Collaboration in Extension Initiatives

Intercultural collaboration in extension initiatives is about bringing together individuals from diverse cultural backgrounds and evaluating the impact and effectiveness of such collaborations on extension programs. It is important to consider several factors to evaluate the impact of intercultural collaboration in extension initiatives. First is the assessment of communication effectiveness. This involves evaluating the clarity and understanding of messages exchanged between individuals from different cultural backgrounds. The effectiveness of intercultural communication can be measured through feedback and observation, ensuring that all participants accurately interpret and understand messages. Second, the evaluation should assess collaboration and cooperation among individuals from different cultural backgrounds. This can be measured by the extent to which individuals work together, share resources, and contribute their unique perspectives and skills to reach common goals. Third, the evaluation should also consider the outcomes and impact of intercultural collaboration on extension programs. This can be measured by examining changes in program participation, knowledge acquisition, behavior change, and overall program outcomes.

## 4.10. The Role of Collaboration in Intercultural Extension Settings

Collaboration is crucial in intercultural extension settings (Blake, 2005). It involves working together across cultural boundaries to achieve common goals, share resources and expertise, and create a sense of collective ownership and responsibility. Collaboration promotes exchanging ideas, perspectives, and best practices, leading to innovative solutions and improved outcomes. In intercultural extension settings, collaboration becomes even more essential as it brings together diverse perspectives and cultural backgrounds, allowing for a more comprehensive understanding of the challenges and needs of the community.

## 4.11. Benefits of Collaboration in Extension

- *Holistic Approach*: Collaboration allows for a holistic approach to agricultural development by integrating multiple perspectives and expertise.
- *Efficiency:* Pooling resources and expertise lead to more efficient and cost-effective implementation of extension activities.
- *Sustainability:* Collaborative efforts promote the sustainability of agricultural practices by ensuring they are culturally appropriate, environmentally friendly, and economically viable.
- *Empowerment:* Collaboration empowers farmers and communities, as they have a voice in decision-making and access to knowledge and resources.

Collaboration in extension work is not just a beneficial option but necessary for achieving meaningful and sustainable impacts. By bringing together diverse stakeholders, extension programs can be more responsive to community needs, more efficient in resource utilization, and more effective in promoting long-term change. The examples and studies presented show that collaboration is not a one-size-fits-all approach but rather a dynamic and adaptable strategy that can be tailored to different contexts and challenges.

## Conclusion

In this chapter, we have delved into the critical realm of intercultural communication and collaboration within the context of extension work. The global agricultural and community development landscape demands a nuanced understanding of diverse cultures, traditions, and communication styles. As highlighted throughout this discussion, effective intercultural communication is not merely a desirable skill but a fundamental necessity for successful extension programs. According to Hall (1959), "Culture is communication, and communication is culture," encapsulating the inherent connection between culture and how we interact within it. Throughout this chapter, we have emphasized several key points. Firstly, cultural competence among extension professionals is paramount. This competence encompasses an awareness of different cultural norms and values and the ability to adapt communication strategies accordingly (Gudykunst & Kim, 2003). Secondly, effective intercultural collaboration involves building trust and relationships across cultural boundaries (Hofstede, 2001). This goes beyond language proficiency to encompass empathy, respect, and an openness to learning from diverse perspectives (Bennett, 1993). Thirdly, while a powerful tool, technology must be used judiciously in intercultural contexts (Jackson, 2017). While it can bridge distances, its effectiveness hinges on understanding cultural nuances.

The importance of effective intercultural communication and collaboration in extension work cannot be overstated. It is the foundation upon which successful partnerships, sustainable development, and impactful outcomes are built. As we navigate an increasingly interconnected world, the ability to communicate and collaborate across cultures is not just a skill; it is a strategic imperative. There are several recommendations for future research and practice in this vital area. Firstly, more studies are needed to explore the most effective training methods for enhancing cultural competence among extension professionals (Chen & Starosta, 2000). These studies should assess long-term impacts on program success and community engagement. Additionally, research focusing on best practices for intercultural collaboration, particularly in the digital age, is crucial (Maznevski & Chudoba, 2000). Understanding how to leverage technology while maintaining cultural sensitivity is a frontier that warrants further exploration. In terms of practical recommendations, extension organizations should prioritize ongoing cultural training and development for their staff (Barker, 2017). This includes immersive experiences, cross-cultural workshops, and mentorship programs. Moreover, platforms for sharing successful intercultural communication strategies and case studies should be established to facilitate knowledge exchange among practitioners globally.

In conclusion, effective intercultural communication and collaboration are not just buzzwords but the bedrock of successful extension works in a diverse world. Extension professionals can make a difference by investing in cultural competence, building meaningful cross-cultural relationships, and harnessing technology wisely. As we move forward, let us remember the words of Nelson Mandela: "If you talk to a man in a language he understands, that goes to his head. If you talk to him in his language, that goes to his heart." The true impact of extension work is realized in reaching the hearts of diverse communities.

## References


Ahmed, S., & Rahman, M. (2020). Building Partnerships for Agricultural Innovation: Lessons from Extension Programs in Pakistan. *Journal of Agricultural Innovation*, 12(1), 24-37.

Anderson, L. M. (2015). Developing Intercultural Competence Through Education and Training. *Journal of International Education Research*, 11(1), 5-14.

Anderson, P., & White, L. (2020). Collaboration as a Key Driver of Extension Program Success. *Journal of Agricultural and Rural Development*, 18(3), 24-37.

Angelelli, C. V. (2004). *Medical Interpreting and Cross-Cultural Communication.* Cambridge University Press.

Barker, G. G. (2017). Building Culturally Competent Organizations. In D. Landis, J. M. Bennett, & M. J. Bennett (Eds.), *Handbook of Intercultural Training* (4th ed., pp. 157-182). Sage Publications.

Bennett, M. J. (1993). Towards Ethnorelativism: A Developmental Model of Intercultural Sensitivity. In R. M. Paige (Ed.), *Education for the Intercultural Experience* (pp. 21-71). Intercultural Press.

Bennett, M. J. (2009). Defining, Measuring, and Facilitating Intercultural Learning: A Conceptual Introduction to the Intercultural Education Double Supplement. *Intercultural Education*, 20(sup1), S1-S13.

Barrett, M., Byram, M., & Lazar, I. (2014). *Developing Intercultural Competence Through Education*.

Biswas-Diener, R., & Diener, E. (2006). The Subjective Well-Being of the Homeless, and Lessons for Happiness. *Social Indicators Research*, 76(2), 185–205.

Brislin, R. W. (2000). *Understanding Culture's Influence on Behavior* (2nd ed.). Harcourt Brace College Publishers.

Chen, G. M., & Starosta, W. J. (2000). Intercultural Communication Competence: A Synthesis. In L. A. Samovar & R. E. Porter (Eds.), *Intercultural Communication: A Reader* (pp. 91-108). Wadsworth.

Chevrier, S. (2003). Understanding the Role of Cultural Dimensions on Technology Acceptance: A Cross-Cultural Analysis of a Model Testing the Impact of Cultural Variations on the Acceptance of a Learning Platform. *Journal of Educational Multimedia and Hypermedia*.

Collier, M. J. (2016). Culturally Responsive Leadership in Community Extension Programs. *Journal of Extension*, 54(1), 1-6.

Cox, T., & Blake, S. (1991). Managing Cultural Diversity: Implications for Organizational Competitiveness. *Academy of Management Executive*, 5(3), 45-58.

Cushner, K., & Brislin, R. (1996). *Improving Intercultural Interactions: Modules for Cross-Cultural Training Programs*. Sage Publications.

Deardorff, D. K. (Ed.). (2009). *The SAGE Handbook of Intercultural Competence*. Sage Publications.

Doe, J., Smith, A., & Johnson, L. (2019). Enhancing Community Engagement Through Collaborative Extension Programs. *Journal of Agricultural Extension*, 25(2), 45-58.

Earley, P. C., & Mosakowski, E. (2004). *Cultural Intelligence*. Harvard Business Review, 82(10), 139–146.

Giles, H., & Ogay, T. (2019). Communication Accommodation Theory. *The International Encyclopedia of Intercultural Communication*, 1-9.

Global Leadership Foundation. (2023). *Cultural Intelligence* (CQ)

Gudykunst, W. B., & Kim, Y. Y. (2003). *Communicating with Strangers: An Approach to Intercultural Communication* (4th ed.). McGraw-Hill.

Gudykunst, W. B., & Kim, Y. Y. (2017). *Communicating with strangers: An approach to intercultural communication*. Routledge.

Hall, E. T. (1959). *The Silent Language*. Doubleday.

Hall, E. T. (1976). *Beyond culture*. Garden City, NY: Anchor Press/Doubleday.

Hall, E. T. (1976). *Beyond culture*. New York, NY: Anchor Books.

Hall, Edward T. (1976). *Beyond Culture*. Anchor Books.

Hallinger, P., & Hord, S. M. (2010). Leading Educational Change: Reflections on the Practice of Instructional and Transformational Leadership. *Journal of Educational Change,* 11(2), 107–134.

Hammer, M. R., Bennett, M. J., & Wiseman, R. (2003). Measuring Intercultural Sensitivity: The Intercultural Development Inventory. *International Journal of Intercultural Relations,* 27(4), 421-443.

Harrell, D., & Bradley, R. (2015). Culture and sustainability in agricultural extension: Perspectives from Honduras. *Journal of Extension*, 53(5), Article 5RIB1. Available online: https://joe.org/joe/2015october/rb1.php.

Hofstede, G. (1980). Culture and organizations. *International studies of management & organization*, 10 (4), 15-41.

Hofstede, G. (1984). *Culture's consequences: International differences in work-related values*. Beverly Hills, CA: Sage Publications.

Hofstede, G. (2001). *Culture's consequences: Comparing values, behaviors, institutions, and organizations across nations*. Sage publications.

Hofstede, G. (2011). *Dimensionalizing cultures: The Hofstede model in context. Online readings in psychology and culture*, 2 (1), 2307-0919.

Hofstede, G., & Bond, M. H. (1988). The Confucius connection: From cultural roots to economic growth. *Organizational Dynamics*, 16(4), 5–21. doi:10.1016/0090-2616(88)90009-.

Hofstede, G., & Minkov, M. (2010). Long-versus short-term orientation: new perspectives. *Asia Pacific Business Review*, 16(4), 493-504.

Hofstede, G., & Minkov, M. (2013). *Values survey module 2013 manual*. York University, Toronto, Canada.

Hong, L., & Page, S. E. (2004). Groups of diverse problem solvers can outperform groups of high-ability problem solvers. *Proceedings of the National Academy of Sciences*, 101(46), 16385-16389.

House, R. J., Hanges, P. J., Javidan, M., Dorfman, P. W., & Gupta, V. (Eds.). (2004). *Culture, leadership, and organizations: The GLOBE study of 62 societies*. Sage Publications.

Hui, C. H., & Triandis, H. C. (1986). Individualism-collectivism: A study of cross-cultural researchers. *Journal of Cross-Cultural Psychology*, 17(2), 225–248. doi:10.1177/0022002186017002006.

Inghilleri, M. (2003). Habitus, field and discourse: Interpreting as a socially situated activity. Target. *International Journal of Translation Studies*, 15(2), 243-268.

Jandt, F. E. (2019). *An introduction to intercultural communication: Identities in a global community*. Sage Publications.

Kim, Y. Y. (2001). Becoming intercultural: An integrative theory of communication and cross-cultural adaptation. *Thousand Oaks*, CA: Sage Publications.

Kim, Y. Y. (2018). *Communication and cross-cultural adaptation: An integrative theory*. Routledge.

LaFasto, F., & Larson, C. (2001). *When teams work best: 6,000 team members and leaders tell what it takes to succeed.* Sage Publications.

Lustig, M. W., & Koester, J. (2013). *Intercultural competence: Interpersonal communication across cultures* ($7^{th}$ ed.). Boston, MA: Pearson.

Lustig, M. W., & Koester, J. (2013). *Intercultural competence: Interpersonal communication across cultures*. Pearson Higher Ed.

Martin, J. N., & Nakayama, T. K. (2019). *Intercultural communication in contexts*. McGraw-Hill Education.

Matsumoto, D., & Hwang, H. C. (2019). Culture and nonverbal behavior. In *The SAGE Handbook of Nonverbal Communication* (pp. 92-110). Sage Publications.

Maznevski, M. L., & Chudoba, K. M. (2000). Bridging Space Over Time: Global Virtual Team Dynamics and Effectiveness. *Organization Science*, 11(5), 473-492.

Minkov, M., & Hofstede, G. (2011). The evolution of Hofstede's doctrine. *Cross Cultural Management: An International Journal*, 18(1), 10–20. doi:10.1108/135276011110 95693.

Ogbimi, E. F. (2018). Cross-cultural communication competence and multicultural team performance in Nigeria. Journal of Intercultural Communication Research, 47(2), 133–156. doi:10.1080/17475759.2018.1452826.

Purkey, W. W., & Schmidt, J. J. (1996). *A handbook for developing multicultural awareness*. Association for Supervision and Curriculum Development.

Ralston, D. A., et al. (2014). Societal-level versus individual-level predictions of ethical behavior: A 48-society study of collectivism and individualism. *Journal of Business Ethics*, 122(2), 283–306. doi:10.1007/s10551-013-1770-7.

Samovar, L. A., Porter, R. E., & McDaniel, E. R. (2019). *Intercultural communication: A reader*. Cengage Learning.

Samovar, L. A., Porter, R. E., McDaniel, E. R., & Roy, C. S. (2020). *Intercultural communication: A reader*. Cengage Learning.

Samovar, L. A., Porter, R. E., McDaniel, E. R., & Roy, C. S. (2020). *Intercultural communication: A reader*. Cengage Learning.

Shuter, R. (Ed.). (2012). *Intercultural new media studies: The next frontier in intercultural communication.* Routledge.

Smith, P. B., & Peterson, M. F. (1988). *Leadership, organizations, and culture: An event management model.* Newbury Park, CA: Sage Publications.

Spitzberg, B. H., & Changnon, G. (2009). Conceptualizing intercultural competence. In Deardorff, D. K. (Ed.), *The SAGE Handbook of Intercultural Competence* (pp. 2-52). Sage Publications.

Taras, V., Kirkman, B. L., & Steel, P. (2010). Examining the impact of culture's consequences: A three-decade, multilevel, meta-analytic review of Hofstede's cultural value dimensions. *Journal of Applied Psychology*, 95(3), 405–439. doi:10.1037/a0018938.

Thomas, D. C., & Peterson, M. F. (2014). Cross-cultural management: Essential concepts (3rd ed.). *Thousand Oaks*, CA: Sage Publications.

Ting-Toomey, S., & Chung, L. C. (2012). *Understanding intercultural communication.* Oxford University Press.

Trompenaars, F., & Hampden-Turner, C. (1998). *Riding the Waves of Culture: Understanding Diversity in Global Business.* Nicholas Brealey Publishing.

Trompenaars, F., & Hampden-Turner, C. (2012). *Riding the waves of culture: Understanding diversity in global business.* Nicholas Brealey Publishing.

Trompenaars, F., & Woolliams, P. (2003). A new framework for managing change across cultures. *Journal of Change Management*, 3(4), 361–375. doi:10.1080/714023845.

Usunier, J. C. (2000). The dimensions of cultural proximity: Differential impacts of sociostructural and cultural attributes on patterns of trade. *Journal of International Marketing*, 8(2), 40–67. doi:10.1509/jimk.8.2.40.19872.

Watzlawick, P., Bavelas, J. B., & Jackson, D. D. (2011). *Pragmatics of human communication: A study of interactional patterns, pathologies, and parado*xes. W. W. Norton & Company.

Wilson, L., & Thompson, S. (2019). Strengthening Agricultural Extension Services Through Collaborative Partnerships. *Journal of Sustainable Agriculture*, 14(2), 78-92.

World Bank. (2018). *Enhancing Agricultural Innovation: How to Go Beyond the Strengthening of Research Systems.* World Bank Group.

## Chapter 5

# Empowering Rural Women: Gender-Responsive Agricultural Extension

**Settipalli Sravani***
**Mathi Girishma**
**Satyapriya**
**and V. Lenin**
Division of Agricultural Extension, ICAR-Indian Agricultural Research Institute, New Delhi, India

### Abstract

Women play a significant role in agriculture, with higher workforce participation rates in rural areas, and act as key stakeholders in many sectors. Despite improvements in education and awareness, gender bias remains a prominent issue in rural areas, leading to socio-economic disparities between men and women. However, their empowerment through technology and education is crucial for meaningful progress. "Agricultural extension programs should address various aspects of rural women's lives, including family, education, health, and cultural activities. To achieve sustainable agricultural development and reduce gender disparities, empowering rural women through gender-responsive agricultural extension is essential." The term "gender-responsive extension" describes agricultural extension services that are developed and offered with consideration for the unique needs and issues that both men and women face in agriculture. It ensures equitable access to resources, information, and opportunities for all women in agriculture and related fields. It involves various strategies, including gender

* Corresponding Author's Email: sravs.harshini@gmail.com

In: Global Horizons in Agricultural Extension Education
Editors: R. K. Doharey, Aman Verma, Satyapriya et al.
ISBN: 979-8-89530-446-4

dynamics, tailoring extension services, capacity building, ensuring access to resources, promoting technology and innovation, fostering partnerships, and monitoring and evaluation. Even though women are vital to the growth of agriculture, they have inadequate access to extension services. A thorough grasp of the significance of gender-responsive extension delivery is necessary to improve this circumstance. This chapter provides an overview of the changes in the extension system in India. Understanding this can improve women's participation in decision-making processes and their livelihoods, ultimately contributing to sustainable agriculture. To bring about revolutionary change, targeted initiatives emphasizing the significance of gender disparities in information access and the establishment of extension on entered-on-led female farmers and extension agents are crucial. Gender-responsive extension services can enhance women's understanding for better decision-making and capacity to implement suitable agricultural practices for crops and animal husbandry. Overall, worldwide gender-responsive agriculture is critical for achieving sustainable development goals related to poverty reduction, food security, and gender equality. It requires a holistic and integrated approach that addresses the structural barriers and inequalities that limit the full participation of women in agricultural development.

**Keywords:** women empowerment, gender-responsive agriculture, agricultural extension systems

## 5.1. Introduction

India's economy heavily relies on its agricultural sector, for 54.6 percent of its workforce is involved in agriculture and allied sectors. Around 80 percent of rural women are engaged in agriculture, making it a crucial source of livelihood for them. Patel and Sethi (2021) reported that among these women, 33 percent work as agricultural laborers, and self-employed farmers account for 48 percent. Rural women catalyze environmental, economic, and social changes in a 'New India' context. Integrating and empowering them in agriculture can drive significant economic growth, enhance food security, and alleviate poverty and hunger.

Women are extensively involved in various agricultural activities such as production, harvesting, and processing. Despite their substantial contribution, rural women face challenges in accessing agricultural information, which hinders their involvement in decision-making regarding technologies, crops, markets, and practices. This lack of access to information limits women's ability to move beyond subsistence farming and hinders their potential to improve their livelihoods. In many cases, women are overlooked in the design of agricultural extension programs, and even when their role is recognized, information flow within households is often biased, with valuable production information not reaching women (Lecoutere et al., 2020).

## 5.2. Gender-Responsive Agricultural Extension

In several developing nations, agricultural extension services are restricted for both genders, yet women are less inclined to utilize these services than men (Meinzen-Dick et al., 2010). The disregard for women farmers is not solely a result of insufficient awareness about gender concerns but also indicates a restricted comprehension of women's requirements, preferences, and ambitions within the agricultural domain. Women comprise 43 percent of the global agricultural workforce, and if they had equal access to resources like technology, financial services, land, markets, and education, and agricultural production could increase, potentially reducing global hunger by 100 to 150 million people (FAO, 2011).

Despite efforts over the past 25 years to focus on women in agriculture, extension services still largely overlook them (Jafry and Sulaiman, 2013). These services are crucial for transferring information, knowledge, and skills through training and demonstrations, but they are often biased toward men, with materials primarily geared toward male members of agricultural households and rarely customized for women. To address these disparities, governments and organizations have sought to increase the presence of female extension and advisory agents, aiming to enhance women's access to services, participation, and agricultural efficiency. Research indicates that women farmers prefer female extension officers (Due et al., 1997), and a higher number of female agents has significantly improved women's productivity and involvement in agriculture (Berger et al., 1984).

Contemporary extension programs acknowledge the importance of involving women but frequently overlook the significant differences in needs between men and women, who are often engaged in diverse agricultural tasks. Additionally, women and men farmers have differing levels of access to resources such as markets, assets, credit, information and other services, all of which are vital for agricultural development. This gender disparity impedes women's productivity, diminishes their impact on the farming sector, and obstructs broader economic and social development goals. (OECD, 2011).

The current approach to supporting rural women in developing nations to attain food security and revenue from sustainable agriculture appears to be lacking. There is a disconnect between what agricultural extension services offer and what women in the farming sector need to improve their livelihoods (Sulaiman et al., 2003).

Gender-responsive agricultural extension is crucial in meeting rural women's specific needs, priorities, and challenges in agriculture. It acknowledges women's diverse roles and aims to equip them with the necessary knowledge, skills, and resources to enhance their productivity, income, and livelihoods. By addressing gender disparities and empowering women, this approach contributes significantly to sustainable agriculture, food security, and rural development. It is indispensable for overcoming these hurdles and unleashing the potential of rural women. This chapter explores the significance of gender-responsive agricultural extension in empowering rural women, examines key strategies and methods, and underscores the advantages of empowering women for sustainable agriculture and rural development.

### 5.2.1. Major Women Empowerment Schemes in India

Since its independence, the government has unveiled numerous programs and flagship schemes to raise rural women's status, providing them with opportunities for a better living and encouraging their participation in paid work. Schemes like National Livelihoods Mission, Pradhan Mantri Kaushal Vikas Yojana (PMKVY), Pradhan Mantri Matru Vandana Yojana (PMMVY), Prime Minister's Employment Generation Program (PMEGP), Beti Bachao Beti Padhao, Deen Dayal Upadhyaya Grameen Kaushalya Yojana (DDU-GKY), among others, have played a significant role in promoting gender equality and the socio-economic empowerment of women in India.

- *Beti Bachao, Beti Padhao (BBBP):* Launched in 2015, this scheme aims to address the declining child sex ratio and promote girls' education.
- *Mahila Shakti Kendra (MSK):* Launched in 2017 under the National Rural Livelihoods Mission, MSK aims to empower rural women

through community participation and access to information, services, and resources.

- *National Mission for Empowerment of Women (NMEW):* This mission aims to empower women by ensuring their economic and social empowerment and addressing issues like violence against women.
- *Pradhan Mantri Matru Vandana Yojana (PMMVY):* This maternity benefit program provides financial assistance to pregnant and lactating women for their first living child.
- *Pradhan Mantri Surakshit Matritva Abhiyan (PMSMA):* This scheme aims to provide comprehensive antenatal care to pregnant women, especially those in rural areas.
- *Ujjawala Scheme*: Launched in 2016, it aims to prevent the trafficking of women and children for commercial sexual exploitation by providing them with rehabilitation services and education.
- *Working Women Hostel Scheme*: This provides safe and affordable accommodation to working women in urban areas.
- *One Stop Centre (OSC) Scheme*: Launched in 2015, this scheme provides integrated support and assistance to women affected by violence in private and public spaces.
- *Women Helpline Scheme (WHS):* Launched in 2016, the WHS provides 24-hour emergency and non-emergency response to women affected by violence.
- *Swadhar Greh Scheme*: This scheme, launched in 2018, provides temporary accommodation, food, and rehabilitation services to women in difficult circumstances.

Currently, rural women have access to various economic benefits, healthcare services, education, productive resources, capacity-building programs, skill development initiatives, and various livelihood opportunities through government programs designed for their benefit.

### 5.2.2. Present Conditions of the World Extension System

This chapter provides an overview of the changes in the extension system in India, Uganda, Ethiopia, and Ghana. An understanding of this can enrich the participation of women in decision-making processes.

#### *5.2.2.1. India*

Since 2005, the ATMA Scheme, also known as the Centrally Sponsored Scheme "Support to State Extension Programs for Extension Reforms," has been in operation. Currently, 691 districts, including 28 states and 5 UTs, are implementing the scheme. The program encourages the nation's decentralized, farmer-friendly extension system. The Scheme's goals are to assist the State Government in its endeavors and provide farmers with access to the most recent agricultural technologies and best practices of various farm themes and related fields through a variety of extension activities, such as Farmer's Training, Farm Schools, Exposure Visits, Farmers Group Mobilization, and Farmers Training.

It serves as a focal point for decentralizing day-to-day operations of the public Agricultural Technology System (ATS) and merging Research and Extension activities. It is a recognized society in charge of district-level technology distribution. As a society, it can accept and use project funding, sign contracts and agreements, and have revolving accounts that might be used to recoup running costs through the collection of fees.

District-level ATMA would eventually lead all district-level technology distribution initiatives. The district's line departments, research institutions, non-governmental organizations, and agencies involved in agricultural development would be connected to it. The primary line Departments of Agriculture, Animal Husbandry, Horticulture, fisheries, ZRS or substations, KVKs, and other research and extension units located within the project districts would join ATMA as constituent members. The institutional identity and affiliation of each Research Extension (R-E) unit would remain intact. Still, the ATMA Governing Board would establish policies and procedures pertaining to district-specific R-E operations, which its Management Committee (MC) would then carry out.

While there is limited evidence on the effects of ATMA on women farmers, specific reports highlight shortcomings in its implementation, with actual allocation and spending falling below the estimated 1.4 million women (about half the population of Nevada) farmers who were supposed to benefit (C. Ragasa, 2014). Lately, there has been widespread recognition of women's participation in agriculture by development agencies. Consequently, significant attention has been paid to this issue at all levels. Women have been included in decision-making processes, ranging from the Governing Board (GB) to the Farmer Advisory Committee (FAC). The practice of nominating 30% non-official women representatives on GBs in ATMA has been adopted. Women's participation at the grassroots level in FAC meetings has been quite encouraging. Women have also occupied numerous different positions in various PIAs. The involvement of non-official women members in decision-making has facilitated women farmers' engagement in various field activities.

### *5.2.2.2. Uganda*

The Ugandan government has previously created and funded several agricultural extension strategies and systems, each with a different need for financial, material, and human resources.

In June 2014, the government decided to fix shortcomings in extension services by restructuring the national agricultural extension system. The Cabinet Sub Committee Report's (2014) recommendations were the foundation for this choice. The "Single Spine Extension System" reforms involved integrating the NAADS program into local government production departments, eliminating parallel institutional structures, and distinguishing the supply of agricultural inputs from the delivery of extension services. Additionally, the reforms involved moving the extension role from the National Agricultural Advisory Services (NAADS) to the core Ministry of Agriculture, Animal Industry and Fisheries (MAAIF) and setting up a Directorate of Agricultural Extension Services (DAES).

The NAADS strategy includes educating district staff on gender issues and identifying indicators to address these issues at the district and sub-county levels. However, its gender impact appears limited. Many individual farmers lack the resources to apply the knowledge acquired, and while NAADS offers extensive training, farmers, especially women, often do not adopt the techniques due to low literacy rates and lack of capital for inputs and technology. Cultural factors also play a role; for instance, some women cannot use certain technologies, like sitting on tractors, due to cultural norms in their communities. Despite the enthusiasm of women and youth to participate in NAADS, they are concerned about the limited range of enterprises available. The enterprise approach of NAADS tends to benefit farming enterprises that require significant land or capital, placing women and youth at a disadvantage compared to men (Bukenya, 2010).

In contrast to men, women, and young people were disadvantaged by the entrepreneurship model that NAADS had adopted, which tended to favor farming operations needing significant quantities of land or cash (Bukenya, 2010). Though many older women are involved in the organizations, few hold leadership roles and thus lack the authority to affect group choices.

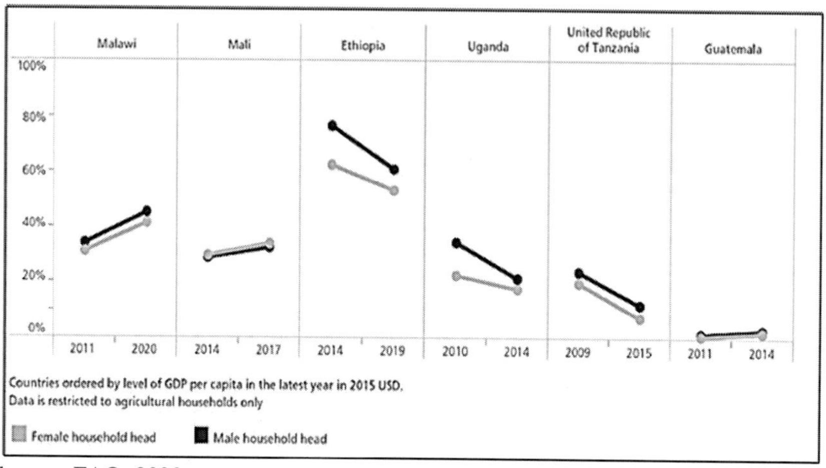


Source: FAO, 2023.

**Figure 5.1.** Women farmers continue to have limited access to extension services compared to men.

### *5.2.2.3. Ethiopia*

In Ethiopia, there is a 23 percent gender productivity gap, partially due to restricted access to extension services designed for women's requirements (Aguilar et al., 2014). Studies indicate that women adopt fewer technologies than men primarily because of limited access to inputs and services (Ragasa, 2012; Manfre et al., 2013).

The Participatory Demonstration and Training Extension System (PADETES), the 'Rural development policies, strategies and instruments document' (FDRE 2001), and a new extension system known as the Participatory Agricultural Extension System (PAES) has been started (MoA 2010). The extension system's human resource and knowledge infrastructure was developed by establishing agricultural technical and vocational education and training (ATVET) and farmer training centers (FTCs). The ATA (Agricultural Transformation Agency) has been instrumental in developing various transformation strategies and testing innovative development programs and approaches.

Gender equality is a national policy priority, and the government has made progress towards recognizing the role of women in the agricultural sector. The National Growth and Transformation Plan (GTP II) identifies women and youth as key cross-cutting influences in all sectoral strategies and development plans (FDRE, 2016). The Ministry of Agriculture has also developed a broader variety of extension packages, recognizing that one size does not fit all farmers. This includes the "women's development package," emphasizing support for women's agricultural activities. Gender equality strategies and mainstreaming guidelines facilitate gender integration in development strategies and programs (MoA, 2011; Mamo and Tagess, 2020). "The gender machinery at all administrative levels serves as a strategic instrument for mainstreaming gender and ensuring its sufficient integration in extension planning, implementation, and evaluation activities."

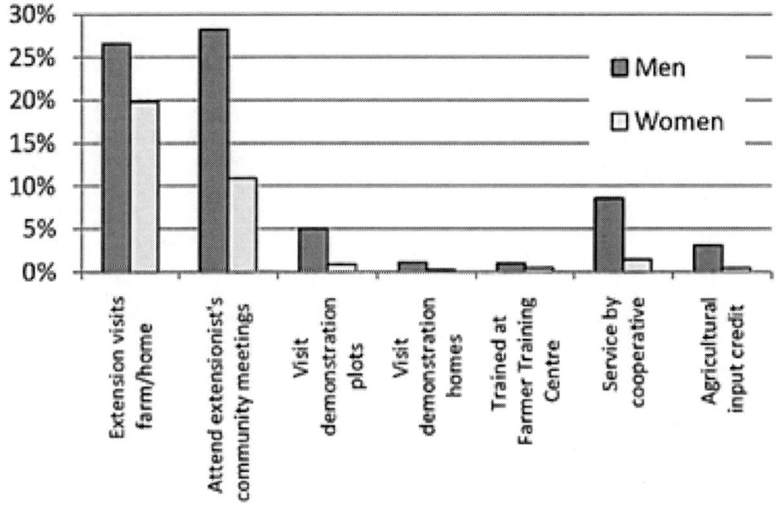


Source: Mogues et al., 2009.

**Figure 5.2.** Percentage of individuals receiving extension and other agricultural services, categorized by gender.

### *5.2.2.4. Ghana*

In Ghana, extension services have historically focused on reaching male farmers, leaving women farmers underserved and marginalized. However, there has been a concerted effort to make extension services more gender-responsive and inclusive in recent years, recognizing women's unique roles and challenges in agriculture.

The Directorate of Agricultural Extension Services (DAES) oversees agricultural technology diffusion by managing extension delivery services in the country. Work with the regional and district administration to ensure that extension services contribute effectively and efficiently towards Ghana's social and economic development. DAES is tasked to perform the following functions:

- Extension Policy formulation and Planning;
- Coordination of Extension activities;
- Provision of Technical support to the regions and districts in the planning and implementation of extension activities;
- Monitoring and Evaluation of all extension activities under the Ministry of Food Agriculture.'

At present, in Ghana, agricultural extension approaches range from the top-down commodity-based approaches to participatory approaches like the Training and Visit (T&V), which was funded by the World Bank, the innovative ICT-based approaches, the farmer field schools (FFSs), commodity participatory approaches, which provide on-line advice to farmers, and the promotion of community radio stations and mobile phones. Over the years, many extension service providers have encouraged these approaches, including producer organizations, farmer organizations, and government and non-governmental organizations (NGOs).

A vital critical initiative to address the gender gap in agriculture extension services is establishing Women in Agriculture Development Units (WIADUs) within the Ministry of Food and Agriculture (MoFA). These units provide specialized support and assistance to women farmers, including training on improved farming practices, access to credit and markets, and information on nutrition and food security. WIADUs aim to empower women farmers and enhance their participation in agricultural development by targeting women-specific needs and priorities.

## 5.3. Benefits for Women Through Gender-Responsive Agricultural Extension

- *Increased Agricultural Productivity*: When women have access to extension services and resources, they can improve their farming practices, leading to better food security and agricultural productivity for their families and communities.
- *Improved Livelihoods:* Empowering women in agriculture can improve their income and livelihoods, leading to better living standards and economic empowerment.
- *Enhanced Food Security*: Women play an essential role in food production and security. Empowering them through extension

services can increase food production and access to nutritious food for their families and communities.

- *Gender Equality*: Gender-responsive agricultural extension services help challenge gender norms and promote gender equality by giving women equal access to information, decision-making opportunities and resources in agriculture.
- *Sustainable Agriculture*: Women often deeply understand local ecosystems and traditional agricultural practices. Empowering them can lead to more sustainable agriculture practices that protect the environment and natural resources.
- *Community Development*: Empowering women in agriculture can lead to broader community development, as women often reinvest their income in their families and communities, leading to improved education, health, and infrastructure.
- *Resilience to Climate Change*: Women are often disproportionately affected by climate change. Empowering them through extension services can help them adapt to climate change and build resilience in their agricultural practices.
- *Improved Nutrition:* Empowering women in agriculture can improve nutrition outcomes because they are often responsible for household activities like food preparation and feeding practices.
- *Social Empowerment*: Empowering women in agriculture can increase social empowerment as women gain confidence, skills, and knowledge that can benefit them both within and outside the agricultural sector.

## 5.4. Challenges Faced by Rural Women in Gender-Responsive Agricultural Extension

- *Limited Resources:* Women often have limited access to land, credit, inputs, and technologies essential for agricultural productivity and income generation. This limits their ability to adopt new practices and technologies promoted through extension services.
- *Time Poverty*: Women in rural areas often bear the burden of household and care work, leaving them with limited time to participate in extension programs or implement new agricultural practices.

- *Limited Mobility and Access to Information:* Women may have limited mobility due to cultural norms or lack of transportation, which restricts their access to agricultural information and extension services.
- *Gender Norms and Discrimination*: Societal norms and gender biases can limit women's participation in decision-making processes related to agriculture and restrict their access to extension services and resources.
- *Lack of Tailored Extension Services*: Extension services often do not consider women farmers' specific needs and constraints, such as their preferred learning methods or the types of crops they cultivate.
- *Financial Constraints*: Women often have limited access to economic resources, which hinders their capacity to invest in agricultural inputs or adopt new technologies.
- *Limited Participation in Agricultural Cooperatives and Groups:* Women may face challenges in joining or participating in agricultural cooperatives or groups, which are often key channels for accessing extension services and resources.
- *Climate Change and Environmental Degradation*: Rural women are disproportionately affected by climate change and environmental degradation, which can impact their agricultural productivity and livelihoods.

## 5.5. Key Strategies for Women's Empowerment Through Gender-Responsive Agricultural Extension

- *Gender-sensitive Training and Capacity Building*: Provide capacity-building programs and training tailored to women farmers' needs and priorities. This includes training on new agricultural practices, technologies, and business skills, as well as on gender issues and the rights of women.
- *Access to Information and Resources*: Ensure that women have equal access to agricultural information, resources, and services. This may include providing information in local languages, using formats that are accessible to women, and ensuring that extension services are geographically accessible to women farmers.

- *Promoting Women's Leadership and Participation*: Encourage and support the leadership of women and contribute to agricultural decision-making processes, including in farmer groups, cooperatives, and community organizations.
- *Financial Inclusion*: Facilitate women's access to economic services, credit, and savings mechanisms, which can help them invest in their farms and businesses.
- *Support for Diverse Livelihoods*: Recognize and support the diverse livelihood strategies of women in agriculture, including their roles in livestock management, natural resource management, and off-farm activities.
- *Addressing Gender Norms and Discrimination*: Challenge harmful gender norms and stereotypes that limit women's participation and decision-making in agriculture. This may include engaging with communities and local leaders to promote gender equality.
- *Tailoring Extension Services*: Ensure that extension services are tailored to the specific needs and constraints of women farmers, considering their roles, responsibilities, and preferred learning styles.
- *Promoting Access to Markets*: Facilitate women's access to markets and value chains by providing training on market access, linking women farmers to markets, and supporting the development of women-led agricultural enterprises.
- *Monitoring and Evaluation*: Monitor and evaluate the impact of gender-responsive agricultural extension programs to ensure that they are effectively meeting the needs of women farmers and empowering them economically and socially.

## Conclusion

Gender-responsive agricultural extension is essential for empowering rural women and unlocking their full potential in agriculture. Recognizing and integrating rural women by safeguarding their access to technology, resources, health services, education, skill development opportunities and ownership rights will improve agricultural productivity and support the creation of an empowered nation. Empowering rural women is the right thing to do and an intelligent investment in the future of agriculture and rural communities.

This chapter examines the various roles of women in agriculture, acknowledging their significant contributions and the challenges they face, including limited access to resources, financial constraints, societal norms, and technological disparities. Despite these challenges, several strategies and programs have emerged to empower women through agricultural extension services. These services not only disseminate knowledge but also promote gender equality by addressing socio-cultural norms and changing perceptions of women's roles in agriculture. Through gender-sensitive approaches, tailored training, and the involvement of female extension workers, these services help reduce gender disparities and create a more inclusive agricultural environment.

Continued efforts are crucial for empowering women in agriculture. This includes policy reforms prioritizing gender-inclusive strategies, increased resources for extension services, improved access to markets and resources for women, and greater participation in decision-making forums. Empowering women in agriculture is not just a development goal but a transformative force that can bring significant societal changes. By harnessing the capabilities of women in agriculture, India can achieve improved food security, economic growth, and social advancement. This empowerment is essential for building a resilient, equitable, and sustainable future for India's agricultural sector and its people. The chapter underscores the ongoing commitment, collaboration, and innovation required to empower women in agriculture, recognizing them as catalysts for transformation and advocates for sustainable development in India's agricultural sector. Achieving fully inclusive and resilient farming systems for the benefit of society demands continuous and targeted investments in women's empowerment.

## References

Aguilar A, Carranza E, Goldstein M, Kilic T, Oseni G. Decomposition of gender differentials in agricultural productivity in Ethiopia. *Policy Research Working Paper Series* 6764, Washington, D.C: The World Bank Group. 2014.

Berger M, DeLancy V, Millencamp A. *Bridging the Gender Gap in Agricultural Extension*, Washington DC, International Center for Research on Women. 1984.

Bukenya C. *Meeting Farmer Demand? An Assessment of Extension Reform in Uganda.* Wageningen University and Research. 2010.

Due JM, Magayane F, Temu AA. Gender again - views of female agricultural extension officers by small holder farmers in Tanzania, *World Development* (1997) 25 (5):713-725.

FAO. *The State of Food and Agriculture 2010-11, Women in Agriculture, Closing the Gender Gap for Development*, Rome, Food and Agriculture Organisation of the United Nations. 2011.

FAO. *Rural Livelihoods Information System (RuLIS).* In: Food and Agriculture Organization of the United Nations. Rome. 2023. https://www.fao.org/in-action/rural-livelihoods-dataset-rulis/en/.

FDRE. *Rural Development Policy, Strategy and Techniques*. Ministry of press and information, Addis Ababa. Ethiopia. 2001.

FDRE. *Growth and Transformation Plan II (GTP II) (2015/16−2019/20)*. Addis Ababa, Ethiopia. 2016.

Jafry T, Sulaiman VR. Gender-sensitive approaches to extension programme design. *Journal of Agricultural Education and Extension* (2013) 19 (5):469-485. https://doi.org/10.1080/1389224X.2013.817345.

Lecoutere E, Spielman DJ, Campenhout BV. Women's empowerment, agricultural extension, and digitalization: Disentangling information and role model effects in rural Uganda. *IFPRI Discussion Paper*, Washington, DC: International Food Policy Research Institute (IFPRI). 2020. https://doi.org/10.2499/p15738coll2.133523.

Mamo H, Tagess L. Gender Differential in Productivity and Income of Haricot Bean in Misrak Badawacho District of Southern Ethiopia. *Ethiopian Journal of Business and Social Science* (2020) 3(1): 42-56.

Manfre C, Rubin D, Allen A, Summerfield G, Colverson K, Akeredolu M. *Reducing the Gender Gap in Agricultural Extension and Advisory Services: How to Find the Best Fit for Men and Women Farmers.* MEAS Discussion Paper 2. 2013.

Meinzen-Dick R, Quisumbing A, Behrman J, Biermayr-Jenzano P, Wilde V, Noordeloos M, Ragassa C, Beintema N. Engendering agricultural research. *Global Conference on Agriculture and Rural Development* (2010) 28–31 March. Montpellier, France.

MoA. *Participatory Agricultural Extension System (Amharic).* Addis Ababa, Ethiopia: Ministry of Agriculture. 2010.

MoA. *Guidelines for Gender Mainstreaming in Agricultural Sector.* Addis Ababa, Ethiopia: Women's Affairs Directorate. 2011.

Mogues T, Cohen MJ, Birner R, Lemma M, Randriamamonjy J, Tadesse F, Paulos Z. *Agricultural Extension in Ethiopia through a Gender and Governance Lens.* Development Strategy and Governance Division, International Food Policy Research Institute – Ethiopia Strategy Support Program 2 (ESSP2) Discussion Paper 007, Ethiopia. 2009.

OECD. *Women's Economic Empowerment; Issues Paper*, DAC Network on Gender Equality (GENDERNET). 2011.

Patel N, Sethi T. Rural Women: Key to New India's Agrarian Revolution. *Kurukshetra*, December (2021). pp. 26-29.

Ragasa C. Gender and institutional dimensions of agricultural technology adoption: A review of literature and synthesis of 35 case studies. *International Association of Agricultural Economists (IAAE) Triennial Conference*, Foz do Iguaçu, Brazil (2012) 18–24.

Ragasa C. Improving gender responsiveness of agricultural extension. *Gender in Agriculture: Closing the Knowledge Gap* (2014) 411-430.

Sulaiman VR, Jafry T, Ashok MS. *Cafeteria for Women in Agriculture*. NCAP Working Paper 4, New Delhi, National Centre for Agricultural Economics and Policy Research. 2003.

# Chapter 6

# Cultivating Resilience: Climate Change and the Future of Extension

**M. Vamshi***
**Siddhesh Sunil Zade**
**Apoorva Veldandi**
**Sanjit Maiti**
**Sanchita Garai**
**and Ramesh Chand Bunkar**
Dairy Extension Division, ICAR-National Dairy Research Institute, Haryana, India

## Abstract

Climate variability is a significant issue that has global implications. It refers to long-term shifts in temperature, precipitation patterns and other climate parameters that could adversely affect the environment, wildlife and human population. Globally, climate change causes adverse effects on agricultural systems, natural resources, livelihood, food and nutritional security, particularly for small and marginal farmers in vulnerable regions. The Indian Network for Climate Change Assessment (INCCA) report warns of impacts such as rising sea level, cyclones, decreased crop produce in rainfed crops, heat stress in livestock, reduced milk productivity and spread of disease. As the repercussions of climate change continue to unfold, there is a need for strategies to enhance resilience within agricultural systems. Many countries depend on their frontline extension systems to cultivate resilience among households. Extension services are pivotal in disseminating knowledge and

* Corresponding Author's Email: vamshi6654@gmail.com

In: Global Horizons in Agricultural Extension Education
Editors: R. K. Doharey, Aman Verma, Satyapriya et al.
ISBN: 979-8-89530-446-4

facilitating adaptation among farmers and stakeholders. This chapter explores the intersection and interaction of climate change and extension services, focusing on cultivating resilience among households from a global perspective, focusing on practical experiences and examining key components of resilience-building within the extension framework. These components include fostering adaptive capacity, promoting sustainable practices, facilitating access to resources and information, and promoting collaboration among diverse stakeholders. The chapter would also shine some light on emerging technologies and innovative approaches that can strengthen extension efforts in addressing climate change. By leveraging extension services effectively, communities can better adjust to the complexities of climatic fluctuations, mitigate risks and build adaptive capacity to ensure sustainable agricultural systems for future generations.

**Keywords:** adaptive capacity, climate change, extension services, resilience

## 6.1. Introduction

Climatic fluctuation is the primary pressing concern, and it has gained attention from scientists, policymakers and the public, who depend directly or indirectly on environment-related activities. Climate change refers to significant and long-lasting alterations in the earth's climate patterns, primarily attributed to anthropogenic activities that release greenhouse gases such as carbon dioxide ($CO_2$), methane ($CH_4$), and nitrous oxide ($N_2O$) into the atmosphere, trapping heat and leading to a gradual rise in global temperatures. As the world warms up, people across the globe are facing new challenges on a scale never experienced before. We are thus living in a world that is climate critical. Solving climate challenges is about reducing the emissions our anthropogenic activities release into the atmosphere and protecting people from harmful climatic events and challenges. Extreme weather events and climate change are becoming more frequent and intense, impacting every country. Still, developing and low-income nations are especially vulnerable to these issues due to their geographic location, reliance on climate-sensitive natural resources and overall development gaps where climate dangers are most likely to occur. If left unchecked, climate risks can reverse years of development in some nations by increasing poverty, escalating inequality, exacerbating food insecurity and causing health concerns.

Additionally, the effects of climate hazards on individuals and communities within nations vary. Ingrained socioeconomic disparities cause most of these effects. Consequently, marginalized groups typically have a significant negative impact on society, which disproportionately exposes them and leaves them vulnerable to climate hazards (UN.org). The 2030 Agenda for Sustainable Development acknowledged that one of the biggest challenges of our time is climate change, whose adverse effects make it difficult for any country to achieve sustainable development. Considering this, the Conference of the Parties to the United Nations Framework Convention on Climate Change (UNFCCC) is recognized as the central forum for negotiating the UNFCCC's global response. The Paris Agreement was adopted on December 12th, 2015, by the Conference of the Parties to the Convention at its twenty-first session. Under this Agreement, the parties committed to quantitatively cutting down their greenhouse gas emissions, which is the prime cause of climatic alterations, and supporting adaptation efforts.

To withstand these challenges and thrive, we need climate resilience. Climate resilience is the ability of a system and its parts to anticipate, absorb, accommodate, and recover from the effects of a hazardous event in a timely and efficient manner, including ensuring the preservation, restoration, or improvement of its essential basic structures and functions (IPCC). It is about successfully coping with and managing the hard climatic impacts on people and preventing them from worsening. This growing interest in resilience building is place-based and interest-based, where communities proactively cohere and develop the ability to adapt to different kinds of shocks and stressors (Berkes & Ross, 2013: Patel et al., 2017). Climate stressors include:

1. Increase in droughts;
2. Floods which damage and destroy crops and adversely affect livestock and shelter;
3. Strong winds and
4. Sea level rise. There has been a rapid growth in resilience-related study in various disciplines and in different contexts and fields.

Through such work, many factors that constrain or enhance resilience at the community and social levels have been identified (Umamaheswari et al., 2021). Diverse income groups and institutions mediating interests and access to resources and opportunities are crucial in forming capacities to withstand shocks and adjust to long-term stressors in rural development and urban studies. (Pandey et al. 2021).

## 6.2. Climate Change Scenario

The United Nations Framework Convention on Climate Change (UNFCCC, 1992) defines climate change as "a change of climate which is attributed directly or indirectly to human activity that alters the composition of the global atmosphere, and which is in addition to natural climate variability observed over comparable periods." These changes are visible in the form of abrupt and anomalous alterations in temperature and rainfall, affecting the weather of numerous vicinities and leading to the alteration of climate-induced disasters experienced worldwide. Global warming, or the increase in average global temperature, is the most evident sign of climatic variations. The global temperature has risen by 0.85°C since 1880 (IPCC, 2014). Climate change has predominant effects on soil, atmosphere, water supply, agriculture, natural ecosystems, and human health—all necessary for the long-term sustainability of most of Earth's production activities (Emeka, 2008). Recent observations show that total greenhouse gas emissions in 2019, including land-use change, reached a new high of 59.1 gigatonnes of carbon dioxide equivalent (Emissions Gap Report, 2020). The WMO report states that 2019 concluded a decade of extreme global heat, melting ice and the highest sea levels due to greenhouse gases produced by human activity, with the global average temperature 1.1°C higher than the pre-industrial period. The average temperatures for the five-year and ten-year periods, 2015-2019, 2010-2019, respectively, are the highest on record. The extent of climatic variation in India, when compared to the global scenario, reveals a greater rate of fluctuations in temperature and rainfall. This variability is expected to escalate in the coming decades, which can create a more gloomy situation in India regarding agricultural production and dairying, which may severely hamper livelihoods. Climate dangers are closely related to India's growth as global warming approaches a tipping point. These hazards seriously jeopardize India's sustainable development since they disproportionately affect weaker, less adaptable communities. In the first nine months of 2022, the country experienced almost daily extreme weather events, ranging from adverse heat and cold waves to cyclones, lightning, floods, and landslides. As a result, 2,755 people lost their lives, over 416,667 houses were destroyed, 1.8 million hectares of cropland were damaged, and about 70,000 animals were killed (Pandey and Sengupta, 2022). India's average temperature has risen by around 0.7°C during 1901–2018, which may increase by 2.4-4.4°C by the end of the 21st century (Krishnan et al., 2020). Summer maximum and winter minimum temperatures showed a rise of 0.5 to 0.9°C, particularly in the Northern region of India. India's susceptibility to droughts has worsened over the past six to seven decades due to a general decline in seasonal summer monsoon rainfall. Between 1951 and 2015, India's summer monsoon precipitation (June to September) decreased by around six percent, with the Indo-Gangetic Plains and the Western Ghats experiencing huge drops (Pandey and Sengupta, 2022). India is the seventh-most vulnerable country concerning climate extremes (Eckstein et al., 2020), wherein five out of twenty Indians are highly vulnerable to all three (flood, drought, and cyclone) extreme events (Pandey et al., 2022).

## 6.3. Effects of Climate Change on the Agriculture and Dairy Sector in India

India's predominant source of income is agriculture, to which dairy and cattle contribute significantly as important enterprises taken up solely or in conjunction with agricultural and other ventures. In addition to providing year-round employment, nourishment, and secondary income, dairying and livestock farming are responsible for the socioeconomic development of both rural and urban communities. Furthermore, India leads the world in milk production, contributing 220 million tonnes, almost 24 percent of global milk production. Dairy products account for nearly one-third of all rural family income in India; 70 million farmers work directly in the dairy business. Milk has a four percent economic share in India and is primarily used for farmers' income (Manohar, 2021).

Climate change adversely affects livestock production in numerous ways, both directly and indirectly. The adverse effects occur in the health and productivity of animals and the production of feed crops and forages. Climate change also directly impacts dairy cattle (Malarkkannan and Kathirchelvan, 2017). Being homeotherms, dairy animals must maintain a narrow range of body temperatures to remain healthy and productive. Stress conditions in animals are caused by ambient temperatures that are either lower or higher than the thermoneutral range. Adult cattle are estimated to operate best between 5°C and 15°C (Hahn, 1999). However, changes in climatic conditions, such as variations in temperature and rainfall, can severely hamper various facets of livestock physiology and health, which may lead to tremendous losses in milk yield and productivity among dairy animals and small ruminants.

## 6.4. Climate Change Adaptation and Resilience

'Adaptation' and 'resilience' are often used interchangeably in policy and academic discourse, but they refer to distinct concepts. Adaptation is the process or action that modifies or changes a system or living organism to enable it to survive in a new environment. In contrast, resilience describes the capacity to respond promptly and efficiently to external pressures and recover from their impacts.

The Intergovernmental Panel on Climate Change (IPCC) defines adaptation as "the process of adjustment to actual or expected climate and its effects." Adaptation discussions often advocate for urgent actions to minimize or avoid harm from climate change before it is no longer feasible. Resilience on the other hand, refers to "the capacity to prepare for, respond to, and recover from the impacts of hazardous climatic events while incurring minimal damage to societal well-being, the economy, and the environment". It also encompasses the magnitude of a disturbance that a system can withstand without crossing a threshold into a new structure or dynamic.

## 6.5. Building Resilience in the Context of Climate Change

To achieve climate resilience and increase sustainable rural livelihood, improving farming systems and reinforcing the non-farm economy is essential (Rathi, 2020). The non-farm economy can increase employment opportunities and income in hazard-prone vulnerable areas. However, the non-farm sector has not received the required support it needs in many non-developing countries (Keshavarz, 2018; Rathi, 2020). As extreme weather events are expected to worsen (IPCC, 2014), enhancing the adaptation of agricultural systems to climate change is necessary. Many countries have introduced several strategies to mitigate climate extremes like reducing cultivation areas of heavy water demanding crops or over-exploiting crops, changing cropping patterns, growing drought-resistant varieties, reducing livestock numbers, and changing ration ingredients (Karimi et al., 2018; Keshavarz and Karami, 2014). Since some of the adaptations are context-specific (IPCC, 2013) and affected by natural, human, and socioeconomic factors more attention should be paid to local realities and socioeconomic contexts of rural people or vulnerable sectors. Some research investigations have recorded how resilience to climate change is highly influenced by household income, ownership, livelihood diversity, place attachment, social cohesion, environmental beliefs, and a utilization of adaptation measures (Alam et al., 2018; Anthopoulou et al., 2017; Birthal et al., 2015; Escarcha et al., 2020; Fang et al., 2018).

## 6.6. A Conceptual Framework for Building Resilience

The concept of resilience has become essential in managing the interaction between human activities and ecosystems sustainably. Historically, ecosystems were often perceived as inexhaustible and self-replenishing, leading to the development of economic tools that exploited them without considering their limits. Agriculture is crucial for producing food, fuel, and fiber, and thus relies heavily on the resilience of both social and ecological systems. Social resilience varies depending on the resources and knowledge available to individuals, communities, and regions as well as the support provided by governments and institutions. Climate, land use, nutrient availability, and the scale of farming operations all have an impact on ecological resilience in agricultural ecosystems. Moreover, agriculture supports the livelihoods of billions of people, particularly those in poverty, thereby contributing directly to societal resilience. To enhance agricultural resilience, policymakers must implement strategies that reduce vulnerability while promoting income generation and poverty alleviation.

## 6.7. Building Resilience by Reducing Vulnerability

Resilience focuses on recognizing how vulnerable society and agriculture are to climate change. It measures the extent of disruption a system can endure without losing its ability to recover or exceed critical thresholds. Therefore, it is crucial to evaluate vulnerability accurately to effectively direct adaptation and mitigation efforts to enhance resilience to climate change. For instance, sensitivity indicators for the agriculture sector may include factors like land resources, input availability, technology, irrigation systems, and production scales. Meanwhile, indicators of adaptive capacity might encompass aspects like rural literacy levels, socioeconomic welfare, and insurance coverage against losses.

### 6.7.1 Actions to Reduce Vulnerability

Vulnerability can be reduced by adopting strategies for mitigation or adaptation to climate change. Institutions and policies—with the advantage of merging or synergy between these strategies—provide a holistic and streamlined path for building resilient communities. Adopting resource-conserving technologies like rainwater harvesting, conservation tillage, INM, and IPM will form a backbone for sustaining and enhancing agroecological systems. An effective policy measure for this would be to promote research and adoption of drought and heat-resistant crop varieties, high genetic makeup, and heat-stress tolerant livestock. Various factors play a significant role in adopting adaptation and mitigation measures, such as the economies of strategies and the benefits or incentives farmers receive when adopting the strategy.

Mitigation measures that help remove and avoid GHG in the atmosphere are another way to build resilience. Reducing the intensity of emissions will decrease the global warming level and thus minimize climatic variability and the need for complex adaptation measures. Many mitigation measures synergize with adaptation strategies by improving ecosystem functioning and enhancing resilience to pests and drought. They are quite economical, motivating farmers to adopt them and help strengthen resilience among rural communities.

### 6.7.2. Outcomes of Improved Resilience

- Increased adaptation of crops and livestock to climate stress
- Improved access to technology and information
- Enhanced income
- Increased renewable energy use, resource-conserving technologies
- Improved risk sharing among communities

### 6.7.3. Resilience, Social Capital and Climate Change

Community resilience is the existence, development, and engagement of community resources by community members to thrive in an environment characterized by change, uncertainty, unpredictability, and surprise (Magis,

2010). With an increased recognition of the potential severity of anthropogenic-led climate change (Goss et al., 2020), there is an increased interest in knowing how local communities and diverse actors can become more resilient to climate-related shocks and stressors. One of the significant subsets related to community resilience building is the role of different social factors that can shape potential or actual collective action for change and resilience (Maclean et al., 2014). Some social factors include local level knowledge sharing, social learning, and good communication among society through the lens of social capital (Mngumi, 2020); these factors all focus on social relationships, networks, and mobilizing collective action within and between different social groups for risk and disaster management responding to various shocks.

### 6.7.4. Relationship Between Social Capital and Community Resilience

Upon reviewing relevant literature on the relationship between social capital and resilience, six key concepts have been identified that contribute to community-level resilience:

1. *Network Quantity:* The number of links among different agents in a social network is believed to enhance social support, trust, and goodwill, which are crucial for improving the ability to respond to various shocks (i.e., reactive resilience) (Cassidy and Barnes, 2012). These studies highlight social networks and family/friends as essential buffers to build resilience.
2. *Network Diversity:* Different types of social capital are essential for addressing immediate crises and identifying areas for improvement (Jordan, 2015). Bonding social capital (interpersonal relationships) is vital for coping with adverse conditions and shocks within communities (Barrett et al., 2011). Bridging social capital (connections across different groups) is necessary for sharing new information and knowledge. This concept assumes that diverse social networks, with rich experience and knowledge are essential for effective responses to future climate impacts.
3. *Balancing Multiple Capitals*: Social capital alone cannot provide complete community resilience. Managing a combination of natural,

physical, economic, and human assets and collective action is crucial for overall resilience.

4. *Role of Formal Institutions:* The contribution of formal institutions to responsive resilience is emphasized. Laws, national policies, regulatory frameworks, and various actors (local government and NGOs) significantly cultivate resilience.
5. *Enhancing Slow-Changing Factors:* This concept highlights the importance of long-term factors like the nature of relationships and natural resources in shaping proactive resilience (Kizos et al., 2014).
6. *Socio-Cultural Dimensions:* This emphasizes the importance of socio-cultural dimensions beyond the help provided by formal institutions or resources in shaping proactive resilience. Community resilience can be proactively developed by overcoming collective norms that exclude certain sections of society and promoting a willingness to change (Bene et al., 2016; Smith et al., 2012).

## 6.8. Role of Extension and Advisory Services in Building Resilience

Delivering extension and advisory services in rural areas across the country presents significant challenges despite their high demand, even without adequate incentives. Despite being under-resourced, these services often struggle with expanding responsibilities placed on their staff without sufficient training. Apart from their traditional role of spreading agricultural technologies, they must address issues stemming from climate variations, disasters, and post-emergency reconstruction. Overcoming resource limitations could enable the extension and advisory services to bolster resilience by serving as a coordinating entity for multiple organizations at different levels during various stages of shocks, owing to their local, subnational, and national connections. The potential roles and methods of extension and advisory services can be viewed through the lens of the sustainable livelihood framework. Overcoming resource limitations could enable the extension and advisory services to bolster resilience by serving as a coordinating entity for multiple organizations at different levels during various stages of shocks, owing to their connections at local, subnational, and national levels. The potential roles and methods of extension and advisory services can be viewed through the lens of the sustainable livelihood framework.

The sustainable-livelihoods framework helps to show how extension and advisory services can build resilience, the framework describes five types of assets/capitals: 1) Human capital, like education, 2) Natural capital, like land and water, 3) financial or economic capital, which includes access to credit, 4) Physical assets like infrastructure, and Social assets like community networks. An asset's value varies based on the community's vulnerability to specific assets. Extension and advisory services can help enhance these assets, including human capital. Several studies suggest that human capital development in the form of education and training for small landholders could be critical for building resilience. In some instances, such as Paine's literature on water management and farming systems in Australia, extension and advisory professionals can help strengthen resilience by improving human capital before utilizing farmers' social capital (networks). They suggest that advisors require good human capital to represent the multiple needs of stakeholders. Some of the potential roles include:

1. Input provision at the time of need and distress
2. Helping the farmers learn how to use new varieties through demonstrations, Farmer Field Scholls
3. Playing the role of an intermediary institution with knowledge of markets and natural resource management
4. Shift from providing technological and management advice to supporting farmers with new skills and information demanded to make informed decisions.
5. Information-sharing tools are mandatory services that climate-vulnerable farmers often require. A farmer who receives quality, up-to-date information and can use it may be able to lessen the effects of climate change.

### 6.8.1. Innovative Models in Promoting Climate Change Adaptation in Agriculture

#### *6.8.1.1. National Innovations in Climate Resilient Agriculture (NICRA)*

NICRA is a platform launched by ICAR, Ministry of Agriculture and Farmers Welfare, GOI in 2011, which brings together all the sectors of agriculture, viz., crops, livestock, horticulture, and fisheries, natural resource management

(NRM), and extension scientists to interact and provide information to meet the challenges faced due to climate change.

*Objectives:*

- To enhance the resilience of Indian agriculture to climatic variability and climate change through strategic research on adaptation and mitigation
- To validate and demonstrate climate-resilient technologies on farmers' fields.
- To strengthen the capacity of scientists and other stakeholders in climate-resilient agriculture
- To draw policy guidelines for broader scale adoption of resilience-enhancing technologies and options

*The main thrust areas covered are:*

- Identifying the most vulnerable districts/regions
- Evolving crop varieties and management practices for adaptation and mitigation,
- Assessing climate change impacts on livestock, fisheries, and poultry and identifying adaptation strategies.
- NICRA demonstrates climate-resilient agriculture technologies through 121 KVKs and technology transfer divisions of institutes to address various climatic extremes and vulnerabilities in the form of four modules.

*The Technology Demonstration Module consists of the following:*

- *Module I:* Natural resource management includes groundwater recharging, conservation tillage, enhanced drainage in flood-prone areas, rainwater gathering and recycling for additional irrigation, and water-saving irrigation techniques.
- *Module II:* Crop Production: The use of drought- and temperature-tolerant varieties, the advancement of rabi crop planting dates in regions experiencing terminal heat stress, water-saving techniques for paddy cultivation (SRI, aerobic, direct seeding), fumigation for temperature management in horticulture, community nurseries for delayed monsoons, customized hiring centers for timely planting, and

location-specific intercropping systems with high sustainable yield index.

- *Module III:* Livestock and Fisheries: Increasing fodder production during droughts and floods, enhancing CPR productivity, promoting better fodder/feed storage techniques, vaccinating against diseases that cause heat stress, managing fishponds and tanks during water scarcity and flooding.
- *Module IV:* Institutional Interventions: New and existing institutional interventions are being strengthened, including those concerning seed banks, fodder banks, custom hiring centers, collective marketing, the introduction of weather index-based insurance, and climate literacy through a weather station at the village level. A Village Climate Risk Management Committee (VCRMC), comprising at least two to three women, is established in each NICRA village to represent all facets of the community and to oversee local-level institutions. Every season's monsoon action plan for NICRA communities is being developed. It primarily addresses emergency plans at the village level. Based on the expected climatic variability, the availability of inputs such as short-duration variety seed and other inputs is guaranteed. The concerned KVKs prioritize technological interventions that enhance farmers' capacity to adapt to changing conditions and increase farming's resilience to climate change. Targeted and appropriate advisories are being created and distributed to farmers promptly to help them deal with unforeseen circumstances. NICRA publishes a monthly e-Newsletter on climate-resilient agriculture and its website as part of its outreach efforts.

#### *6.8.1.2. Village Knowledge Center — ClimaAdapt*

The Village Knowledge Center (VKC) serves as a channel for the dissemination of information, expertise, and skills to rural communities by utilizing the most appropriate Information and Communication Technologies (ICTs). With support from the Norwegian Embassy and NIBIO, the M.S. Swaminathan Research Foundation (MSSRF) has constructed eight VKCs for the ClimaAdapt initiative. To help the community make informed decisions, the VKC aims to provide a sustainable single-window knowledge platform. Farmers can access location-specific climate-smart agro advisories and technologies, and the community will receive demand-responsive, comprehensive information and knowledge tailored to their specific knowledge needs.

#### *6.8.1.3. HARITA-PRIYA*

- A Wireless Sensor Networks (WSN)- a disease forewarning and crop advisory model
- HARITA (Harmonised Information of Agriculture, Revenue, and Irrigation for a Transformation Agenda) and PRIYA (Precision Technology for Agriculture) develop a model that will acquire micro-climatic information from farmers' fields using WSN technology and enable the agriculture department to formulate and disseminate personalized advisory.

#### *6.8.1.4. e-Arik (e-Agriculture)*

To observe crop conditions and to diagnose pests, diseases, and deficiencies at the field level. Digitally document these issues using ICT and communicate the information to staff at the e-Arik station to get recommendations back to the village.

## 6.9. Climate Smart Agriculture Practices/Technologies

1. *Direct-seeded rice:* In traditional rice farming, rice is first sprouted in a nursery before the seedlings are moved onto a heavily tilled area with standing water. Direct-seeded rice is sown directly in a dry seedbed like any other upland crop. This dramatically lowers the amount of water required by the crop, does away with the tedious procedure of physically transplanting seedlings, and enhances the physical characteristics of the soil.
2. *Alternate wetting and drying in rice*: Rice fields that undergo alternate flooding and draining are subjected to alternate wetting and drying. Farmers can make more informed decisions about when to irrigate their fields by using monitoring tools such as tensiometers. Compared to continuous flooding, alternate wet and drying reduces methane emissions by an average of 48%. Using this in conjunction with precision fertilizer tools will help cut greenhouse gas emissions even more.

3. *ICT services to access weather and agro advisories*: M-Solution is an ICT-based climate and agro advice initiative financed by CCAFS that CIMMYT is currently implementing in Climate-Smart Villages in collaboration with IFFCO Kisan Sanchar Limited (IKSL) as a content partner and Kisan Sanchar as the implementation partner. Farmers receive voice and text messages with information on new seed varieties, climate-smart farming techniques, weather forecasts, and conservation agriculture advice. The project's objectives are to record farmers' perspectives on increasingly unpredictable weather events and determine whether the information they get influences their behavior in general to help them adopt new techniques and technologies and adapt to climate change. The project made sure that women farmers were included from the beginning. Many have stated that it is an essential source of climate change knowledge for them.
4. *Zero-tillage:* Growing crops without tillage—without disturbing the soil—is known as zero-till or no-till farming. It improves the retention of organic matter, the cycling of nutrients in the soil, and the quantity of water that percolates into the soil. Zero-tillage enhances the qualities of the soil, increasing its resilience. According to a CCAFS-CIMMYT study conducted in Haryana, zero-tillage benefits the environment and the economy. According to the results, farmers who choose zero-tillage over conventional tillage can save almost USD 79 per hectare in overall production expenses and improve net revenue by roughly USD 97.5 per hectare. According to the study, converting to a zero-tillage wheat production method from conventional tillage lowers greenhouse gas emissions by 1.5 mg CO2 equivalent per hectare.
5. *Laser land leveling*: A tractor-towed, laser-controlled tool that produces a very smooth surface is called a laser-leveller. Raising the field's level guarantees that water is distributed and reached equally, which boosts crop yield. Since there is less need to run electric pumps when there is less water, it also improves energy efficiency. In Haryana, most farmers rent the machinery on an hourly basis. Farmers pay an hourly rate of 600–700 Rupees for the use of the machines. In the state, laser land levelling has become an increasingly popular procedure. At the present adoption size, Haryana is expected to save 933 million cubic meters of irrigation water annually. An estimated 163,600 MT of CO2 equivalent is mitigated annually from greenhouse gases. (CIMMYT-CCAFS, 2014).

6. *Residue management/mulching*: The practice of keeping a protective layer of stubble and vegetative remains on the soil's surface is known as crop residue mulching. It enhances the amount of organic matter in the soil, enhancing the seedbed's quality and the soil's ability to absorb and hold water. In Haryana, burning rice harvest residue is one of the main problems. Crops can be drilled directly without tillage using cutting-edge planting equipment like the Turbo Happy Seeder, with the residue left on the surface acting as mulch.
7. *Crop diversification*: Increasing a farmer's crop mix is the goal of crop diversification, which helps them become less reliant on any crop for income. This increases household food security and diversifies a farmer's exposure to climate risk. Additionally, opportunities for diversification based on system optimization and an inventive crop portfolio support sustainable intensification—legume additions to the crop rotation cycle aid in fixing nitrogen in the soil.
8. *Agroforestry:* Planting trees beside vegetables or crops aids in soil erosion prevention and carbon sequestration. In addition to giving crops shade, trees yield fruit, wood, fodder, and fuel for farmers. They improve biodiversity by offering homes to various bird, insect, and animal species and supporting a robust biological landscape.

## 6.10. Case Studies and Best Practices

Extension agencies and other stakeholders cannot underestimate the importance of significantly improving the resilience of agricultural communities. Moreover, comprehending the potential impacts of such interventions can provide valuable insights for future climate-focused projects, particularly in aligning with the needs of smallholders and commercial farmers. This understanding lays a robust foundation for effective policy advocacy to comprehensively enhance agricultural commodities' resilience. Therefore, this section explores various interventions geared towards strengthening resilience. Moreover, comprehending the potential impacts of such interventions can provide valuable insights for future climate-focused projects, particularly in aligning with the needs of smallholders and commercial farmers. This understanding lays a robust foundation for effective policy advocacy to comprehensively enhance agricultural commodities' resilience. Therefore, this section explores various interventions geared towards strengthening resilience.

Chaudhari and Kendall (2020) undertook an intervention wherein the experts and the villagers were motivated to collaborate and work together to build a community resilient to the impacts of climate change. In this attempt, ICT tools, namely Information systems, were incorporated in numerous villages, giving engendered information suitable for both genders with distinct agricultural roles. The information system provided the timely rainfall forecasts and temperature data, which, after farmers' experience on traditional knowledge, were utilized in sound decision-making regarding numerous agricultural practices. It focussed on three aspects of rural agriculture: activities for withstanding the effects of climate change, activities for managing resources efficiently, and finally, replacing the maladaptation practices with climate-smart ones. Thus, a collaborative approach was followed to improve the agricultural communities' adaptive, absorptive, and transformative capacities, which form the preamble for climatic resilience efficiently and finally replace the maladaptation practices with climate-smart ones. Garg *et al.* (2020) conducted research in the central Bundelkhand region of India wherein climate-smart practices concerning water management were incorporated in the villages using extension agents. The practice led to an improved level of groundwater table by 2-5 meters, which led to relief of the community from water scarcity. This led to increased area being brought under cultivation. The 20 percent of land hitherto fallow was brought under intensive cultivation. Further, crop yields increased from 11 to 60 percent, leading to a rise in household income.

A study by Šūmane et al., (2018) underscored the utility of various attempts at addressing the issues of building resilience, which were carried out by the RETHINK project. An extension approach in Austria demonstrated the significance of the farmer-led networks in leveraging the working of vocational training schools and field schools concerning agriculture. These networks focussed on informal knowledge building and transfer of knowledge. Further, a case study conducted in Denmark focused on a watershed wherein the farmers learned local landscape management, resulting in an enhanced level of understanding at individual and community levels. Further, this attempt led to the enhancement of social capital development.

Further, in Ireland, farmers worked as mediators to disseminate novel strategies to apply organic manure. The case focused primarily on novel nutrient management methods. The farmers utilized decision support systems aided with collaborations in the form of a network. This led to improved utilization of nutrients. Saran et al., (2022) devised an approach called PROTOCOL, which included numerous agricultural interventions tailored explicitly for women farmers and aimed at improving climate resilience. It highlighted the incredible impact of building women farmers' adaptive capacity via numerous approaches, namely knowledge dissemination, finance-based natural resources management, and social networking. These led to positive outcomes such as monetary increment, socio-demographic alleviation, and improvement mitigation and decision-making.

## 6.11. Lessons Learned from Resilience Initiatives

Based on numerous case studies undertaken, it was comprehended that extension agencies are important for enhancing the delivery of numerous interventions to make farmers climate-smart, thereby leading to the build up of resilience. As detailed in Table 6.1, various national and international organizations have developed initiatives to support climate resilience in agriculture. Extension-based interventions attempt bilateral learning between numerous stakeholders and the farming community to formulate a sound knowledge dissemination network (Brooks & Loevinsohn, 2011). Such networks not only aid in gaining timely information but also improve the probability of making the communities self-sufficient with regards to the generation of a sound knowledge base from where they can procure all needed information for the sustainability of livelihood and agricultural practices. The case studies reveal that extension agents play a significant role in demonstrating and creating awareness about various climate-smart activities among farmers. The attempts at the development of knowledge networks lead to better decision-making and better collaboration among the farmers. Thus, implementing climate-smart interventions at the village level requires proper work from the extension workers. Utilizing ICT can improve the implementation by making the farmers self-sufficient in terms of required knowledge and regular update of their knowledge regarding agricultural practices, demand and supply of various farm commodities, financing, etc. Extension agents have the scope to work in the direction of enhancing collaboration among numerous stakeholders (Faure et al., 2016). Further, the extension agents can effectively persuade multiple stakeholders to build cohesiveness with the farming communities and thereby promote the development of organisations which can enhance the productivity, profitability and marketing in the agricultural enterprises be it farmers producer organisations, farmer producers' companies, or certain public-private relationships Darr et al., 2014).

**Table 6.1.** List of Initiatives by some National and International organizations

| S.No | Agency/Organisation | | Initiatives |
|---|---|---|---|
| 1 | Government of India | National Action Plan on Climate Change | • National Mission on Sustainable Habitat, National Water Mission, National Mission for Sustainable Agriculture, Green India Mission and National Mission for sustaining the Himalayan Ecosystem, National Mission on strategic knowledge for climate change<br>• Pradhan Mantri Kisan Urja Suraksha Evem Utthan Mahabhiyan (PM KUSUM)<br>• Jal Jeevan Mission<br>• Prime Minister Krishi Sinchayee Yojana<br>• Rashtriya Krishi Vikas Yojana |
| 2 | International Initiatives on Climate Change | World bank | • Action Plan on Climate Change Adaptation and Resilience<br>• Maharashtra Project on Climate Resilient Agriculture<br>• India National Cyclone Risk Mitigation Project<br>• Sustainable Livelihoods and Adaptation to climate change<br>• Tamil Nadu and Puducherry Coastal Disaster Risk Reduction Project<br>• Bihar Kosi Basin Development Project |
| | | UNDP | • Project on Enhancing climate resilience of India's coastal communities |
| | | ADB | • Project on Strengthening climate change resilience in urban India |
| | | UNFCCC | • Global Environment Facility<br>• Adaptation Fund<br>• Green Climate Fund |
| 3 | Civil society initiatives | | • Community-based Microclimate resilience<br>• Community-Based flood early warning system<br>• Building Agriculture Resilience by Harvesting rainwater in India<br>• Pathways to climate-resilient livelihood models |

**Table 6.1.** (Continued)

| S.No | Agency/Organisation | | Initiatives |
|---|---|---|---|
| 4 | Private sector initiatives | BASIX and ICICI, Lombard | • Microinsurance reduces farmers' exposure to weather risk |
| | | ITC Limited | • Adaptation to climate change impact through diversification of farming systems |
| | | Pepsico India | • Replenishing Water |
| | | Tata International Limited, TCS | • mKRISHI: Empowering Rural Farmers |
| | | Fasiam Agro Farms and Jammu & Kashmir Medicinal & Aromatic Plants | • Adapting to climate change by growing medicinal and Aromatic Plants |

## 6.12. Policy Directions for Strengthening Extension and Advisory Services for Improved Resilience

Various institutes working on the mandate of mitigating the negative impacts of climate change make recommendations for formulation of policies such that those policies mitigate the adverse effects on all the vulnerable communities – especially the poor, marginal or women community – through suggestions in many sectors such as agriculture, water resources, biodiversity, and disaster management. These policies have come a long way in enhancing the resilience of communities. However, these policies need to be further altered to address the newer challenges that climate change presents. First, such policies should promote sustainable agriculture. These policies should promote climate-resilient crop varieties and water-efficient irrigation techniques, such as agroforestry (Aishwarya & Kumar, 2024; De Costa, 2020). Moreover, providing weather information and insurance schemes to farmers can help reduce risks from climatic variability and extreme weather events. Second, these institutes must emphasize the conservation and sustainable utilization of water resources for all. Suggestions for harvesting rainwater, groundwater availability, and watershed management programs should be made.

NICRA, an institute in India dedicated to combating the effects of climate change, focuses on equipping regions with climate-resilient water supply infrastructure to ensure access to potable water in dry areas. It also works to strengthen the resilience of biodiversity conservation programs through activities such as expanding protected areas, restoring degraded ecosystems, and protecting biodiversity hotspots and endangered species. Additionally, NICRA aims to enhance its ecosystem-based adaptation strategies by incorporating emerging climate science into contingency plans. Furthermore, it prioritizes disaster risk reduction and management programs, including strengthening early warning systems, improving the resilience of infrastructure, and enhancing community preparedness and response mechanisms.

Policies that bring communities and local governments together to plan for effective disaster risk reduction and climate change mainstreaming in urban planning and development processes (such as promoting green infrastructure, sustainable transportation, and energy-efficient buildings) would contribute to reducing carbon emissions and building urban climate resilience. Few policy recommendations have put NICRA on the road to achieving climate-resilient development for the many. But, given that climatic variability is multidimensional, policy shifts must move beyond NICRA and address the different facets of climate vulnerability. Bringing to the forefront issues of sustainable agriculture, water resource management, biodiversity conservation, disaster risk reduction, and urban resilience, NICRA is responsible for guiding communities toward a resilient future that can weather/reduce the effects of the impending climate impacts. It is of utmost necessity that NICRA must be equipped with climate-resilient water supply infrastructure to ensure potable water in dry areas, build up the resilience of its biodiversity conservation programs wherein undertaking of numerous activities like: expanding the protected area estate, restoring degraded ecosystems, protecting biodiversity hotspots and enhancement of disaster risk reduction and management programs like strengthening early warning systems, increasing the resilience of infrastructure, and enhancing community preparedness and response mechanisms should also be undertaken.

### 6.12.1. Policy Directions for Strengthening Extension and Advisory Services Towards the Promotion of Resilience

Policy guidelines aimed at strengthening extension and advisory services to bolster resilience play a role in tackling the challenges brought about by climatic challenges and biotic stressors that lead to significant impacts. Therefore, policymakers must craft recommendations that prioritize enhancing the capacity reach and effectiveness of extension services to empower farmers and communities to adapt to evolving conditions and foster resilience. Below are some policy directions:

- *Inclusive and Participatory Approaches:* It is imperative to ensure that a variety of stakeholders, such as small-scale farmers, female farmers, young individuals, and indigenous groups, are considered when developing inclusive and participatory extension service approaches. Involving communities in decision-making processes and adaptation strategies can enhance the effectiveness and relevance of extension services (IFAD, 2017).
- *Integration of Indigenous Knowledge:* Policy measures should encourage the integration of farming methods and indigenous wisdom into extension services. Indigenous communities often

possess insights and adaptive techniques that complement knowledge, leading to more culturally relevant and context-specific solutions for enhancing resilience.

The policies must promote and harness the immense capabilities of information and communication technologies (ICT) and digital solutions in delivering extension services. Such systems might comprise mobile applications, online portals, and multimedia resources allowing farmers to obtain data such as weather forecasts, market information, good agricultural practices, and virtual advice (World Bank, 2017).

- *Collaboration and Partnerships:* Policies should be articulated to promote partnership and cooperation between government agencies, civil society organizations, research institutions, and the private sector, which will help strengthen extension services. Providing resources such as expertise, networks, and connections in various fields by different stakeholder's aids in scaling up and customizing extension services efficiently (UNDP, 2018).
- *Monitoring and Evaluation:* Policies should be made to support intense monitoring and evaluation systems so that the impact and efficacy of extension services in enhancing resilience can be clearly identified. These should include indicators such as adoption rates for climate-smart practices, changes in productivity, income gains, and improvements in community resilience related to climate-induced shocks and stresses (IFPRI, 2016).

## 6.13. Emerging Trends in Climate Change Research and Extension

Numerous research works can be sought which have focused on the various facets of climate change. Numerous types of research focus on climatic parameters with respect to their variability across specific time slices. In contrast, many go one step further by correlating these variabilities with the effects on various walks of life, be it agriculture, livestock, livelihood, etc (Perez et al., 2015; Suryanto & Rahman, 2019). As far as social sciences-based research is concerned, numerous literature sources can be found that have thoroughly studied the impact of climate-smart interventions on specific social strata such as marginal farmers, smallholder farmers, women, etc. (Gbetibouo et al., 2010; Srinivasa Rao et al., 2016). Recently, there has been a growing trend in assessing resilience and vulnerability scenarios, with various frameworks employed for this purpose (Negi et al., 2012; Philpott et al., 2008). This surge has fueled ongoing advancements in developing new frameworks and methodologies for evaluating the impacts of climate change. Within extension research, diverse studies have explored the effects of demonstrations on society's adaptive capacities and investigated socio-psychological factors such as perception, attitude, and knowledge of households regarding climate change. Additionally, research has focused on the adoption of practices to enhance sustainability in the face of climate change, the impact of utilizing agro-advisory services on farmers, experimental studies on interventions' effects on adaptive capacity, and the documentation of indigenous knowledge for mitigating climate change effects. Consequently, a wide range of studies is underway in the field of climate change.

## 6.14. Leveraging Innovation for Resilience

Despite considerable efforts by numerous institutions to address climate change through adaptation and mitigation measures, the impact of these interventions has been relatively limited. Therefore, there is a pressing need to develop innovations that surpass previous attempts in effectiveness, efficiency, sustainability, and socioeconomic upliftment, particularly for impoverished populations more vulnerable to climate change effects. Sarkodie et al., (2023) emphasized that strategies for reducing global emissions may necessitate radical changes in energy consumption and production, which are fundamental to economic structures. As a developing country, India faces significant energy demands due to factors like population growth, urbanization, and infrastructure expansion (Avtar et al., n.d.; Madlener & Sunak, 2011), posing a dual challenge of mitigating climate change while sustaining development. Therefore, energy innovations have the potential to significantly enhance pathways to climate resilience, particularly through transitions to renewable energy sources. Moreover, innovations in carbon abatement technologies, such as using clean fuels, can play a crucial role in mitigating the effects of climate change (Sarkodie et al., 2023). Furthermore, adaptive governance can be an effective tool by promoting innovative approaches in policy-making that incorporate experimentation and social learning. For example, in Uganda, adopting governance replaced the public extension services system with a private contact-based system (Cooper & Wheeler, 2015), facilitating learning responses to societal and environmental feedback. This approach fosters knowledge sharing and encourages innovation to address uncertainties arising from climate change, ultimately enhancing the resilience of small-scale farmers through alternative livelihoods.

## 6.15. Challenges in Climate Extension Services

Even though significant efforts are made by various institutions to combat climate change through the agency of extension services, be it private or public, the extension services find great hindrance in disseminating knowledge regarding climate-smart activities. A study in Uganda highlighted the lack of manpower, poor organizational structure, and motivational problems as major bottlenecks in disseminating climate-centric information to farmers. Moreover, numerous extension agents find hindrances in disseminating extension services due to difficulty in transportation, which arises from poor road connectivity in numerous areas and a lack of funds. Further, a dearth is found in quality extension material, leading to ineffective dissemination or increased complexity of the subject matter as perceived by the clientele system (Chinseu et al., 2019). Further, more excellent farmer-to-agricultural extension agent ratios burden extension agents already highly overburdened with work. Further, a lack of funds for expenditure on implementing climate change adaptation practices was prevalent in numerous countries (Bryan et al., 2009). The major hindrance is resistance towards change on the part of the farmer community, which shows skepticism towards accepting novel innovations and technologies, which can lead to better resilience for the farmers (Meijer et al., 2015).

## Conclusion

Climate change has garnered significant global attention due to its profound impact on the livelihoods of a large portion of the population, especially small and marginal farmers. With over 50 percent of the rural population reliant on the agricultural sector, which is susceptible to various hazards, there is an urgent need to implement comprehensive policies incorporating adaptation and mitigation measures into projects. Extension services play a crucial role in achieving these targets at the grassroots level, enhancing the ability to cope with climate extremes. In summary, the case studies and best practices discussed above underscore the importance of efforts by extension agencies and related stakeholders to enhance the resilience of agricultural communities against climate change threats. Through a range of interventions, including the use of ICT tools, informational campaigns, adoption of climate-smart practices, and inclusive approaches tailored to the needs of vulnerable minorities, significant success can be achieved in enabling farmers to adapt to new conditions. However, numerous challenges persist, such as resource constraints, inadequate infrastructure, and resistance to change, necessitating concerted efforts to strengthen extension services through policy-making initiatives.

The way forward would be to continue to promote climate resilience locally, nationally, and internationally through the three existing levels of grassroots initiatives, policy advocacy, and global cooperation. Policies should be designed to support an inclusive and participatory approach to support the integration of indigenous knowledge, ensure full utilization of technology, promote collaboration, and create effective monitoring and evaluation mechanisms. Moreover, alongside opportunities focused on people, involvement in energy transitions, innovative governance, and transforming livelihoods will lay the groundwork for a more resilient and sustainable agricultural sector. To conclude, by tackling these challenges and leveraging these opportunities, I am confident that we can realize an integrated agriculture sector capable of withstanding the effects of climate change and ensuring adequate food production for future generations.

## References

Aishwarya, S., Kumar, P. (2024). A Perspective Way to Climate Smart Agriculture. *Technological Approaches for Climate Smart Agriculture*, 9–24. https://doi.org/10.1007/978-3-031-52708-1_2.

Alam, G. M. M., Alam, K., Mushtaq, S., Filho, W. L. (2018). How do climate change and associated hazards impact the resilience of riparian rural communities in Bangladesh? Policy implications for livelihood development. *Environmental Science*. 84, 7–18.

Anthopoulou, T., Kaberis, N., Petrou, M. (2017). Aspects and experiences of crisis in rural Greece. Narratives of rural resilience. *Journal of Rural Studies*. 52, 1–11.

Avtar, R., Tripathi,S., Aggarwal, A., Resources, P. K. (2019). Population–urbanization–energy nexus: a review. *Mdpi.Com*. Retrieved April 10, 2024, from https://www.mdpi.com/2079-9276/8/3/136.

Balasubramani, N., Dixit, S. (2016). Innovative extension models in promoting climate change adaptation in agriculture. Manage

Barrett, G., Vanderplaat, M., Gonzalez, M. E. C., Irmao, J. F., Ampuero, M. C. G., Vera, C. E. M. (2011). Civic networks and community resilience in Brazil, Canada, Chile, and Cuba. *Journal of Civil Society*. 7: 333–362.

Bene, C., Al-Hassan, R. M., Amarasinghe, O., Fong, P., Ocran, J., Onumah, E., Ratuniata, R., Tuyen, T. V. (2016). Is resilience socially constructed? Empirical evidence from Fiji, Ghana, Sri Lanka, and Vietnam. *Global Environmental Change*. 38: 153–170.

Berkes, F., Ross, H. (2013). Community resilience: Toward an integrated approach. *Society and Natural Resources*, 26(1), 5-20.

Birthal, P. S., Negi, D. S., Khan, M. T., Agarwal, S. (2015). Is Indian agriculture becoming resilience to droughts? Evidence from rice production system. *Food Policy*. 56, 1–12.

Brooks, S., Loevinsohn ,M. (2011). Shaping agricultural innovation systems responsive to food insecurity and climate change. *Natural Resources Forum*, *35*(3), 185–200. https://doi.org/10.1111/j.1477-8947.2011.01396.x.

Bryan, E., Deressa, T. T., Gbetibouo, G. A., Ringler, C. (2009). Adaptation to climate change in Ethiopia and South Africa: Options and constraints. Environ. Sci. Policy 12 (4), 413–426. https://doi.org/10.1016/j.envsci.2008.11.002.

Cassidy, L., Barnes, G.D. (2012). Understanding household connectivity and resilience in marginal rural communities through social network analysis in the village of Habu, Botswana. *Ecology and Society*. 17: 18.

Chinseu, E., Dougill, A., Stringer, L. (2019). Why do smallholder farmers dis-adopt conservation agriculture? Insights from Malawi. Land Degradation. Dev. 30 (5), 533–543.

Cooper, S. J., Wheeler, T. (2015). Adaptive governance: Livelihood innovation for climate resilience in Uganda. *Geoforum*, *65*, 96–107. https://doi.org/10.1016/j.geoforum.2015.07.015.

Darr, D., Hoffmann, V., Helmle, S. (2014). *Extension Services for Rural Development*. 205–240. https://doi.org/10.1007/978-3-642-41404-6_8.

De Costa, W. A. J. M. (2020). Increasing Climate Resilience of Cropping Systems in Sri Lanka. *Agricultural Research for Sustainable Food Systems in Sri Lanka: Volume 2: A Pursuit for Advancements*, *2*, 107–157. https://doi.org/10.1007/978-981-15-3673-1_6.

Eckstein, D., Winges, M., Kunzel, V., Schafer, L., Korperschaft. (2020). Global Climate Risk Index 2020 Who Suffers Most from Extreme Weather Events.

Emeka, D. O. (2008). Impact of climate change on livelihood sustainability in the lake chad region of Nigeria in Popoola. Proceedings of the 32nd *Annual Conference of Forestry Association of Nigeria.*

Escarcha, J., Lassa, J. A., Palacpac, E. P., Zander, K. K. (2020). Livelihoods transformation and climate change adaptation: the case of smallholder water buffalo farmers in the Philippines. *Environmental Development*. 33, 100468.

Fang, Y., Zhu F., Qiu X., Zhao, S. (20180. Effects of natural disasters on livelihood resilience of rural residents in Sichuan. *Habitat International*. 76, 19–28.

FAO. (2015). Transforming rural extension: New challenges and approaches. Food and Agriculture Organization of the United Nations.

Faure, G., Davis, K., Ragasa, C., Franzel, S., Babu, S. (2016). Framework to Assess Performance and Impact of Pluralistic Agricultural Extension Systems: The Best-fit Framework Revisited. In *IFPRI Discussion Paper 01567*. https://doi.org/10.13140/RG.2.2.10224.05129.

Garg, K. K., Singh, R., Anantha, K. H., Singh, A. K., Akuraju, V. R., Barron, J., Dev ,I., Tewari ,R. K., Wani, S. P., Dhyani, S. K., Dixit S. (2020). Building climate resilience in degraded agricultural landscapes through water management: A case study of Bundelkhand region, Central India. *Journal of Hydrology*, *591*. https://doi.org/10.1016/j.jhydrol.2020.125592.

Gbetibouo, G. A., Ringler, C., Hassan, R., (2010). Vulnerability of the South African farming sector to climate change and variability: An indicator approach. *Natural Resources Forum*, *34*(3), 175–187. https://doi.org/10.1111/j.1477-8947.2010.01302.x.

Goss, M., Swain, D. L., Abatzoglou, J. T., Sarhadi, A., Kolden, C. A., Williams, A. P., Diffenbaugh, N. S. (2020). Climate change is increasing the likelihood of extreme autumn wildfire conditions across California. *Environmental Research Letters* 15: 094016.

Hahn, G. L. (1999). Dynamic responses of cattle to thermal heat loads. *Journal of Animal Science*, *77*, https://doi.org/10.2527/1997.77suppl_210x.

IFAD. (2017). Enabling the rural poor to overcome poverty. International Fund for Agricultural Development.

IFPRI. (2016). Adaptation and resilience to climate change in low-income countries: A global assessment. International Food Policy Research Institute.

IPCC. (2013). *Climate change 2013: the physical science basis.* In: Stocker, T. F., Qin, D., Plattner, G. K., Tignor, M., Allen, S. K., Boschung, J., Nauels, A., Xia, Y., Bex, V., Midgley, P. M. (Eds.), Contribution of Working Group I to the Fifth Assessment Report of the Intergovernmental Panel on Climate Change. Cambridge University Press, Cambridge, United Kingdom and New York, NY, USA, p. 1552.

IPCC. (2014). Summary for policymakers. Edenhofer, O., Pichs-Madruga, R., Sokona, Y., Farahani, E., Kadner, S., Seyboth, K., Adler, A., Baum, I., Brunner, S., Eickemeier, P., Kriemann, B., Savolainen, J., Schl¨omer, S., von Stechow, C., Zwickel, T., Minx, J. C. (Eds.), *Climate Change 2014: Mitigation of Climate Change*. Contribution of Working Group III to the Fifth Assessment Report of the Intergovernmental Panel on Climate Change. Cambridge University Press, Cambridge, United Kingdom and New York, NY, USA.

Jordan, J. C. (2015). Swimming alone, the role of social capital in enhancing local resilience to climate stress: A case study from Bangladesh. *Climate and Development*. 7: 110–123.

Karimi, V., Karami, E., Keshavarz, M. (2018). Vulnerability and adaptation of livestock producers to climate variability and change. *Rangeland Ecology and Management*. 71 (2), 175–184.

Keshavarz, M. (2018). Addressing barriers of rural development under drought. *Journal of Research*. Rural Plan. 7 (2), 97–118.

Kizos, T. V., Detsis, T., Iosifides, Metaxakis, M. (2014). Social capital and social-ecological resilience in the Asteroussia Mountains, Southern Crete, Greece. *Ecology and Society* 19: 11.

Krishnan, R., Sanjay, J., Gnanaseelan, C., Mujumdar, M., Kulkarni, A., Chakraborty, S. (2020). *Assessment of Climate Change over the Indian Region*. Springer. https://doi.org/10.1007/978-981-15-4327-2.

Maclean, K., Cuthill, M., Ross, H. (2014). Six attributes of social resilience. *Journal of Environmental Planning and Management*. 57: 144–156.

Madan, N., Rishta, P. Bhandari, R., Sharma,V. (2022). *Climate Resilience*. New Delhi:Shakti Sustainable Energy Foundation.

Madlener, R., Sunak, Y., (2011). Impacts of urbanization on urban structures and energy demand: What can we learn for urban energy planning and urbanization management? *Sustainable Cities and Society*, *1*(1), 45–53. https://doi.org/10.1016/j.scs.2010.08.006.

Magis, K. (2010). Community resilience: An indicator of social sustainability. *Society and Natural Resources*. 23: 401–416.

Malarkkannan, S. P., Kathirchelvan, M. (2017). Effect of climate change and adaptation measures in dairy industry of Kerala. *International Journal of Applied and Pure Science and Agriculture*, *3*(1), 139–146.

Manohar, A. (2021). *The Indian Dairy Landscape*. https://www.investindia.gov.in/team-india-blogs/indian-dairy-landscape

Meijer, S. S., Catacutan, D., Ajayi, O. C., Sileshi, G. W., Nieuwenhuis, M., (2015). The role of knowledge, attitudes and perceptions in the uptake of agricultural and agroforestry innovations among smallholder farmers in sub-Saharan Africa. *Int. J. Agr. Sustain* 13 (1), 40–54. https://doi.org/10.1080/14735903.2014.912493.

Mngumi, L. E. (2020). Exploring the contribution of social capital in building resilience for climate change effects in peri-urban areas, Tanzania. *GeoJournal.* 86: 2671–2689.

Negi, G. C. S., Samal, P. K., Kuniyal, J. C., Kothyari, B. P., Sharma, R. K., Dhyani, P. P. (2012). Impact of climate change on the western Himalayan mountain ecosystems: An overview. *Tropical Ecology*, *53*(3), 345–356.

Pandey, A., Prakash, A., Werners, S. E. (2021). Matches, mismatches and priorities of pathways from a climate-resilient development perspective in the mountains of Nepal. *Environmental Science & Policy*, *125*, 135-145.

Pandey, K., Sengupta R. (2022). *An assessment of extreme weather events*.

Patel, S. S., Rogers, M. B., Amlot, R., Rubin, G. J. (2017). What do we mean by community resilience. A systematic literature review of how it is defined in the literature. *PLoS currents*.

Perez, C., Jones, E. M., Kristjanson, P., Cramer, L., Thornton, P. K., Förch W., & Barahona C.(2015). How resilient are farming households and communities to a changing climate in Africa? A gender-based perspective. *Global Environmental Change*,*34*,95–107. https://doi.org/10.1016/j.gloenvcha.2015.06.003.

Philpott, S. M., Lin, B. B., Jha, S., Brines, S. J., (2008). A multi-scale assessment of hurricane impacts on agricultural landscapes based on land use and topographic features. *Agriculture, Ecosystems and Environment*, *128*(1–2), 12–20. https://doi.org/10.1016/j.agee.2008.04.016.

Rathi, A. (2020). Is agrarian resilience limited to agriculture? Investigating the "farm" and "non-farm" processes of agriculture resilience in the rural. *Journal of Rural studies*. https://doi. org/10.1016/j.jrurstud.2019.12.015.

Saran, A., Singh, S., Gupta, N., Walke, S. C., Rao, R., Simiyu, C., Malhotra, S., Mishra, A., Puskur, R., Masset, E., White, H., Sharma, Waddington, H. (2022). PROTOCOL:Interventions promoting resilience through climate-smart agricultural practices for women farmers: A systematic review. *Campbell Systematic Reviews*, *18*(3). https://doi.org/10.1002/cl2.1274.

Sarkodie, S. A., Ahmed, M. Y., Owusu, P. A., (2023). Advancing COP26 climate goals: Leveraging energy innovation, governance readiness, and socioeconomic factors for enhanced climate resilience and sustainability. *Journal of Cleaner Production*, *431*, 139757. https://doi.org/10.1016/j.jclepro.2023.139757.

Smith, J. W., Anderson, D. H., Moore R. (2012a). Social capital, place meanings, and perceived resilience to climate change. *Rural Sociology*. 77: 380–407.

Srinivasa, C., Gopinath, K. A., Prasad, J. V. N. S., Prasannakumar., Singh A. K. (2016). Climate Resilient Villages for Sustainable Food Security in Tropical India: Concept, Process, Technologies, Institutions, and Impacts. *Advances in Agronomy*, *140*, 101–214. https://doi.org/10.1016/bs.agron.2016.06.003.

Šūmane, S., Kunda, I., Knickel, K., Strauss, A., Tisenkopfs, T., Rios, I., des I., Rivera M.,Chebach, T., Ashkenazy A., (2018). Local and farmers' knowledge matters! How integrating informal and formal knowledge enhances sustainable and resilient agriculture. *Journal of Rural Studies*, *59*, 232–241. https://doi.org/10.1016/j.jrurstud.2017.01.020.

Suryanto, S., Rahman, A., (2019). Application of livelihood vulnerability index to assess risks for farmers in the Sukoharjo regency and Klaten regency, Indonesia. *Jamba: Journal of Disaster Risk Studies*, *11*(1). https://doi.org/10.4102/jamba.v11i1.739.

Umamaheswari, T., Sugumar, G., Krishnan, P., Ananthan P. S., Anand A., Jeevamani J. J. J., Rao C. S. (2021). Vulnerability assessment of coastal fishing communities for building resilience and adaptation: Evidence from Tamil Nadu, India. *Environmental Science & Policy*, *123*, 114-130.

UNDP. (2018). Accelerating climate-resilient and low-carbon development: The Africa–Nordic Climate Solutions Pitch Hub.

What is the difference between climate change adaptation and resilience? - Grantham Research Institute on climate change and the environment. (2024). Grantham Research Institute on Climate Change and the Environment. Retrieved October 17, 2024 from https://www.lse.ac.uk/granthaminstitute/explainers/what-is-the-difference-between-climate-change-adaptation-and-resilience/.

World Bank., (2017). Digital technologies for agricultural and rural development. World Bank Group.

## Chapter 7

# Evaluating the Impact and Measuring the Success of Extension Programs: A Global Perspective

**Chethan N. D. Patil[1,*]**
**Sapna Jarial[1]**
**Rahul Dundesh Bellagi[2]**
**and Dayanand Patil[3]**
[1]Department of Agricultural Economics and Extension, School of Agriculture, Lovely Professional University. Phagwara, Punjab, India
[2]Research Associate, Indian Council of Social Science Research, Centre for Multi-disciplinary Development Research, Dharwad, Karnataka, India
[3]Division of Dairy Extension, ICAR-National Dairy Research Institute, Karnal, Haryana, India

## Abstract

This chapter delves into the crucial processes of evaluating the impact and measuring the success of extension programs. It emphasizes the need to move beyond measuring outputs and delve deeper into the program's influence on target audiences. It also outlines a framework for practical evaluation, outlining key considerations such as clearly defined objectives, targeted impact indicators, and appropriate data collection methods. It explores various evaluation approaches that assess the achievement of program objectives and participatory models that incorporate stakeholder feedback. Key themes explored include defining impact indicators, selecting appropriate impact evaluation

[*] Corresponding Author's Email: chethanpatilnvl@gmail.com

In: Global Horizons in Agricultural Extension Education
Editors: R. K. Doharey, Aman Verma, Satyapriya et al.
ISBN: 979-8-89530-446-4

methodologies, and addressing resource constraints and stakeholder engagement challenges. Challenges associated with extension evaluation are addressed, including isolating the program's specific impact from external factors and ensuring data collection methods are feasible and cost-effective. The chapter emphasizes tailoring evaluation strategies to each extension program's unique context and goals. By providing practical guidance and best practices, this chapter empowers extension professionals to design robust evaluation plans demonstrating the program's effectiveness in fostering knowledge acquisition, behavior change, and positive outcomes for target audiences. Evaluation approaches that assess the achievement of program objectives and participatory models that incorporate stakeholder feedback. Key themes explored include defining impact indicators, selecting appropriate impact evaluation methodologies, and addressing resource constraints and stakeholder engagement challenges. Challenges associated with extension evaluation are addressed, including isolating the program's specific impact from external factors and ensuring data collection methods are feasible and cost-effective. The chapter emphasizes tailoring evaluation strategies to each extension program's unique context and goals. By providing practical guidance and best practices, this chapter empowers extension professionals to design robust evaluation plans demonstrating the program's effectiveness in fostering knowledge acquisition, behavior change, and positive outcomes for target audiences.

**Keywords:** impact, evaluation, impact evaluation, extension programs

## 7.1. Introduction

Evaluation of extension programs involves systematically gathering information about extension activities, their characteristics, and the program outcomes to make judgments for improving effectiveness and decision-making for future programs. The demand for evaluating extension programs in India is demand-driven and based on government policies. Earlier, during the green revolution phase, the emphasis was on production and, later, on productivity. Many factors are responsible for making extension evaluation an essential area of concern. The paradigm shifts in extension policies, budget crunches, and a sense of accountability for each penny spent thus contribute to a vital role expected from evaluation (Paxson & Schady, 2002). There are specific questions, which are expected from the funders of extension programs, such as:

1. How was the money utilized?
2. Why should the extension program be funded?
3. What is the effect of the programs?
4. How will ineffective program be identified and improved/ discontinued?

Extension Evaluation is the only answer to all these questions.

### 7.1.1. What is Extension Program Evaluation?

Scriven (1991) states that evaluation is all about determining the merit or worth of the object being evaluated, be it a program, project, product, policy or event. Evaluation is a systematic, rigorous, and meticulous application of scientific methods to assess a program's design, implementation, improvement, or outcomes (Ross et al., 2004). Thus, Extension Evaluation can be defined as an assessment of the worth or value of a programs design, implementation, improvement or outcome systematically using scientific methods. There are different forms of evaluation: Stake (1973) classified evaluation based on methodology; Worthen et al., (1997) emphasized the position of the significant audience; and Owen (1993) categorized evaluation based on purpose. Since Owen (1993) attempted to classify field-based evaluation studies, the meta-model of Owen can be used to classify evaluation approaches into five types based on the purpose of evaluation:

- Impact Assessment
- Programs Management
- Process Evaluation
- Design Clarification
- Programs Development

### 7.1.2. What Are We Evaluating?

In general, evaluation assesses an extension program's progress, outputs, outcomes, and impact. A few commonly used analytical aspects described by Dale (2004) are given below:

1. *Relevance:* To what extent the programs or initiative has addressed, or is addressing, high-priority issues as seen by current and potential stakeholders, particularly beneficiaries
2. *Effectiveness:* How much are the desired impacts (development goals), projected changes, planned outputs, and intended effects (immediate and effect objectives) being created or achieved?
3. *Impact:* The overall effect of any program or initiative on the target beneficiaries and any additional participants
4. *Efficiency:* The number of outputs produced, and the quality of the resources invested (money and human efforts)
5. *Sustainability:* The preservation or enhancement of favourable outcomes brought forth by the assessed project or programs (or its component)
6. *Replicability:* The program's or project's viability to be repeated, in whole or in part, in a different setting (i.e., later, for a distinct population, by a different organization, etc.)

Development programs and policies generally aim to bring about positive changes in areas like income, education, and health. However, public policy discussions often overlook whether the changes are accomplished. Instead, program managers and policymakers tend to focus on tracking and reporting the program's resources and immediate results, such as the amount of funding allocated, the distribution of textbooks, or the number of individuals participating in job training. Evaluating whether programs have successfully achieved their desired outcomes of improving overall conditions is less commonly prioritized.

Impact evaluations play a crucial role in the shift towards evidence-based policymaking on a global scale. This shift emphasizes focusing on outcomes and results rather than just inputs, ultimately reshaping public policy. The emphasis on results monitoring national and international targets and aids program managers in enhancing accountability, determining budget allocations, and guiding program design and policy decisions. Monitoring and evaluation are essential components of evidence-based policymaking, providing stakeholders with tools to assess and enhance policies and programs' quality, efficiency, and effectiveness. This focus on results is vital at both the program management level, where decisions on cost-effectiveness and budget allocations are made, and at the country level, where ministries compete for funding (Newman et al., 2002). Additionally, evidence serves as a foundation for transparency and accountability, ensuring that governments are accountable to their citizens for the performance of public programs.

The robust evidence produced by impact evaluations is increasingly becoming the cornerstone for enhanced accountability, innovation, and learning. In a landscape where policymakers and civil society are pressing for results and accountability in public programs, impact evaluation can offer robust and credible evidence on performance and, importantly, on whether a specific program has achieved or is achieving its intended outcomes. Impact evaluations are also progressively utilized to experiment with new ideas in program design or service delivery. On a global scale, impact evaluations are crucial in advancing knowledge about the effectiveness of development programs by shedding light on what works and what doesn't in reducing poverty and enhancing welfare.

An impact evaluation evaluates changes in individuals' well-being that can be linked to a particular project, program, or policy. This emphasis on attribution is the distinguishing feature of impact evaluations. Consequently, the primary challenge in conducting effective impact evaluations is identifying the causal relationship between the program or policy and the outcomes of interest.

## 7.2. Impact Assessment

The impact measurement involves evaluating the positive and negative, primary and secondary long-term effects that result from an intervention, whether they are intended or unintended, and whether they occur directly or indirectly. Assessing the impact of an intervention on the user group is a crucial aspect of evaluating the effectiveness of a technological intervention. The World Bank defines impact assessment to determine whether the program has achieved its desired effects (both positive and negative) on individuals, households, and institutions and whether these effects can be attributed to the program intervention (Baker, 2000). Impact assessment (I.A.) is vital in program planning, implementation, and evaluation. I.A., along with monitoring and evaluation, is utilized to ensure that the program progresses in line with its objectives and verify whether the implementing agency has achieved the desired results. In the context of development projects, impact assessment is relatively straightforward, as extensive literature, experience, and accepted norms and practices. However, this is not the case in the humanitarian sector. The rationale for conducting impact assessments in the humanitarian sector industry stems from adopting "Results Management" techniques and the belief that impact assessment practices. However, the lack of agreed-upon norms and standards across the humanitarian sector remains a significant challenge in conducting impact assessments.

### 7.2.1. Meaning of Impact Assessment

Impact Assessment is a comprehensive tool that measures the effectiveness of organizational activities and evaluates the significance of changes brought about by those activities. It is a unique blend of art and science, as it requires analytical skills and creativity. Impact assessment is closely tied to the organization's mission and has a ripple effect throughout the entire organization. The ability to assess and articulate impact is a powerful means of communication. Impact evaluation is just one of the many approaches that support evidence-based policymaking. It encompasses various methods, including monitoring and other types of assessment.

Monitoring is an ongoing process that continuously tracks the progress and developments within a program. It utilizes collected data to inform program implementation, management, and decision-making. The primary focus of monitoring is to track financial disbursement and program performance against expected results and trends over time. Monitoring is essential to all programs and provides critical information about program performance, including implementation and costs. Typically, monitoring tracks inputs, activities, and outputs but can also include outcomes, such as progress toward achieving national development goals.

Evaluations are periodic, objective assessments of a planned, ongoing, or completed project, program, or policy. They are utilized selectively to address specific design, implementation, and results questions. Unlike continuous monitoring, evaluations are conducted at distinct time points and often involve seeking an external perspective from technical experts. The design, methodology, and cost of evaluations vary significantly based on the question being addressed. Evaluations can tackle three types of questions:

- Descriptive questions include the current situation, focusing on processes, conditions, organizational dynamics, and stakeholder perspectives.

- Normative questions compare the current situation to the ideal scenario, evaluating activities and achieving targets. Normative questions can pertain to inputs, activities, and outputs.
- Cause-and-effect questions delve into attribution, exploring the impact of interventions on outcomes.

### 7.2.2. Prospective versus Retrospective Impact Evaluation

Impact evaluations can be categorized into two main types: prospective and retrospective. Prospective evaluations are developed concurrently with the program design and are integrated into the program implementation process. Baseline data is gathered before the program is executed for the treatment group (receiving the intervention) and the comparison group (not receiving the intervention). On the other hand, retrospective evaluations of the program's impact post-implementation, examining treatment and comparison group's ex-post (Pitt & Khandker, 1998).

Prospective impact evaluations are more likely to yield robust and credible evaluation findings due to several reasons. Firstly, collecting baseline data allows for establishing outcome measures before the program commences. Baseline data is crucial for comparing groups and ensuring their similarity. It also aids in evaluating the targeting effectiveness of the program, determining whether it reaches its intended beneficiaries.

Secondly, defining success measures during the program planning phase focuses on the program itself and the evaluation of its intended outcomes. Impact evaluations are rooted in a program's theory of change or results chain. The evaluation design helps in clarifying program objectives by establishing clear success measures. Policymakers must develop precise program goals and evaluation questions to ensure the results are highly relevant to policy decisions. The support of policymakers is crucial for a successful evaluation, as it provides the legitimacy and value of the review in informing important policy decisions. Lastly, in prospective assessment, the treatment and comparison groups are identified before the evaluation of the intervention takes place.

## 7.3. The Framework of Impact Assessment

Impact assessment (I.A.) plays a crucial role in the planning, execution, and evaluation of programs. With monitoring and evaluation, I.A. is utilized to ensure that the work progresses in line with the program objectives and to determine whether the achieved results align with the desired outcomes set by the implementing agency. I.A. measures the attainment of project milestones, outputs, and outcomes and their influence on the targeted population. The effectiveness of impact assessment relies on the effectiveness of the program implementation process.

The concept of impact revolves around measuring the changes brought about, and impact assessment aims to establish a cause-and-effect relationship between the inputs and the magnitude or scale of these changes. Figure 7.1 depicts the framework as a foundation for integrating impact assessment with monitoring and evaluation. Impact assessment evaluates the contribution of outputs and outcomes towards the purpose and overall goal of the program. Typically, the impact itself is not explicitly incorporated within the framework. Figure 1 illustrates the process involved in conducting an impact assessment.

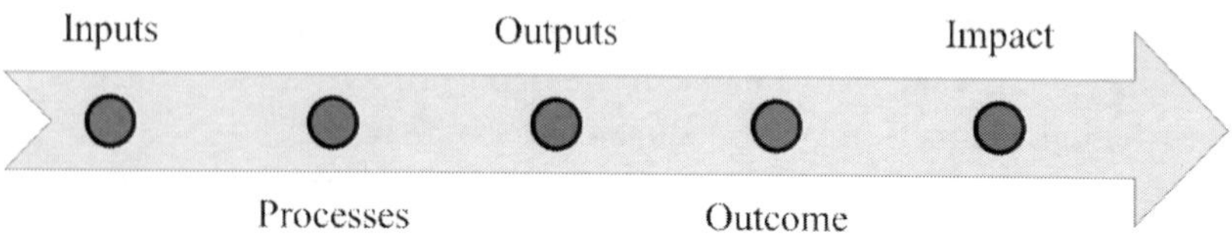


**Figure 7.1.** Impact Assessment Framework.

### 7.3.1. Elements of the Impact Assessment Framework

- *Inputs* – All the resources that contribute to producing and delivering outputs. Inputs include finances, personnel, equipment, and buildings
- *Outputs* – The finished items and services that are prepared for delivery
- *Outcomes* – Outcomes represent the medium-term effects on specific beneficiaries resulting from attaining outputs, aligning with an institution's strategic goals and objectives outlined in its plans.

- *Impacts* – The long-term effects of reaching goals, such as employment creation and poverty reduction.

## 7.4. Impact Evaluation in Agricultural Extension

"Impact" is a commonly used term in extension, yet its precise definition is not widely understood. Blakenburg (1995) defines impact as enduring and sustainable changes brought about by a particular intervention in the lives of beneficiaries. The impact can encompass both expected and unexpected changes resulting from an extension intervention, which may be linked to the program's specific objectives. It is important to note that impact can manifest in both positive and negative forms. Therefore, impact assessment of extension programs involves examining any outcomes or effects of an extension activity on individuals' lives, whether intentional or unintentional or tangible or intangible. Dart et al. (1998) identified four approaches to impact evaluation in Agricultural Extension:

- *Goal-based Impact Evaluation*
  This form of impact evaluation assesses whether a program has achieved its stated goals or objectives. It involves developing valid measures of goals.
- *Need-based Impact Evaluation*
  In certain situations, goal-based evaluation may have limitations, leading to the need for goal-free assessment. Scriven (1967) introduced goal-free evaluation, arguing that evaluating only a program's goals is insufficient for determining its societal value. This gave rise to needs-based assessment, which focuses on assessing society's needs rather than solely measuring the extent to which a program's goals have been met.
- *Financial Analysis of Outcome*
  Financial analysis of an Extension outcome can be conducted using various methods such as Cost-Benefit Analysis (C.B.A.), Net Present Value (N.P.V.), Cost-Effectiveness Analysis (C.E.A.), Additional Economic Returns (A.E.R.), Partial Budgeting Technique (PBT), and Internal Rate of Return (IRR). This impact evaluation can effectively determine whether farmers' income has increased over a specific period.

- *Impact Evaluation for Illumination*
  Impact evaluation for lighting aims to shed light on an extension program's impact by analyzing the intended and unforeseen effects. Impact evaluation is a type of valuation that focuses mainly on the welfare of communities, households, and individuals due to interventions, program, initiatives, and policy measures (reforms). Appropriate methodological answers are necessary for assessing the effects of policies on their settings and levels. The impact of these interventions can serve as a framework for discussing the outcomes. These two issues underscore the importance of evaluating the extent of impact, which remains a significant concern."

### 7.4.1. Impact of What?

What are the independent variable interventions we are looking for?

- Training
- Technological intervention
- A developmental program

### 7.4.2. Impact on What?

This matter pertains to the 'dependent variable' dilemma. Interventions often influence various organizations, groups, and individuals. What level of impact should be of interest to us? Given this diversity, it is beneficial to differentiate between two primary levels of impact for scoping purposes: the institutional and beneficiary levels. This encompasses various effects across all stages of the results chain, including indirect consequences on farm households, communities, institutions, technical or social systems, and the environment. After outlining this distinction, it is crucial to recognize that for many individuals in the development sector, impact assessment revolves around the impact at the beneficiary level. The primary focus is on how policy interventions, either directly or indirectly, impact the well-being of beneficiaries and to what degree changes in well-being can be linked to these interventions.

### 7.4.3. Agree on What Is Valued

Achieving a practical impact evaluation necessitates striking a harmonious equilibrium between acknowledging the values held by stakeholders and considering the intricate empirical nature of the transformative processes brought about by an intervention. Within extension program and projects, both intended and unintended consequences arise, frequently translated into quantifiable indicators as early as the design phase. Furthermore, there exists a dichotomy between short-term and long-term effects. While specific interventions yield specificity impacts, others may require a significantly longer duration to manifest their effects, which may evolve. By focusing on short- or intermediate-term outcomes, we can gain insights into the significance of implementing designs that can measure effects over the long run.

### 7.4.4. How Are Impact Indicators Chosen? (Adapted from Badioli, 2011)

The selection of impact indicators plays a vital role in the evaluation process of a program or intervention, as they offer a quantifiable means to measure the success and outcomes of the initiative (Heckman & Todd, 1998). As illustrated in Figure 7.2, the process of selecting indicators for assessing impacts of technology or extension interventions involves the following systematic steps. The following steps outline a systematic approach to choosing impact indicators:

- Determine the level of impact assessment required, whether it be output, outcome, or impact;
- Identify a concise list of relevant indicators that effectively capture the chosen level of impact;
- Determine the data sources accessible and the method of data collection necessary for each indicator;
- Develop a matrix that outlines the indicators, highlighting their significance in relation to the selected impact level (high/low), the ease of data collection (easy/feasible but effort/challenging levels), and the associated cost;

- Prioritize the indicators based on importance, ease of data collection, and cost, utilizing a weighted matrix ranking/rating system to select a set of indicators;
- Categorize the chosen indicators by the data source to identify the sources capable of providing data on clusters of indicators;
- Finalize the selection of indicators and establish a data collection plan based on the available logistical, human, financial resources and time constraints.

| | |
|---|---|
| **Social and behavioural indicators** | Changes in |
| | Awareness, knowledge, and skills level of beneficiaries; Policy changes |
| | Behaviour, e.g., adoption |
| | Community capacity |
| | Organisational capacity (skills, structures, resources) |
| | Increases in service usage |
| **Socio-cultural indicators** | Food security |
| | Nutritional implications. |
| | Reduction in poverty |
| | Status of women |
| | Variations in resource allocation |
| | Changes in cash requirement |
| | Labour distribution |
| | Distribution of benefits across gender and geographical locations |
| **Technology level indicators** | Adoption of improved technology – symbolic and actual adoption; |
| | Horizontal impact expansion in the cultivation of superior variety/breed |
| | Vertical impact – rise in the yield of superior variety/breed |
| | Decrease in production expenses |
| | Mitigation of risks |
| | Growth in yearly earnings/financial capability |
| | Employment opportunities generated |
| **Environmental impact measures** | Impact on soil erosion and degradation, sedimentation, soil compaction, soil pollution, and water pollution |
| | Alterations in hydrological pattern |
| | Impacts on biodiversity, atmospheric pollution, and greenhouse gas emissions |
| **Institutional impact measures** | Modifications in the organizational framework |
| | Shift in the quantity of researchers |
| | Variations in the makeup of the research group |
| | Utilization of interdisciplinary methods and enhancements |
| | Adjustments in the financial resources assigned to the initiative |
| | Modifications in the involvement of both public and private sectors |
| | Introduction of novel techniques or methodologies |

**Figure 7.2.** Indicators for Assessing Impacts of Technology or Extension Interventions.

### 7.4.5. Indicators for Assessing the Impacts of Technology or Extension Interventions

Assessing the impacts of technology or extension interventions requires careful selection of indicators that capture the outcomes and effects of the intervention on various stakeholders, systems, and environments.

- *"Technical Aspects":* Physical input-output of goods and services;
- *"Institutional Aspects":* Organizational and managerial factors, such as customs, ownership, local organizations, and cultural contexts;
- *"Socio-cultural Aspects":* Wider social consequences, distribution of resources and income, and employment prospects;
- *"Commercial Aspects":* Business and financial considerations, ensuring a steady supply, and meeting market demand;
- *"Economic Aspects":* Economic efficiency, costs and benefits;
- *"Environmental Aspects":* Biological and physical effects.

### 7.4.6. Quantitative Impact Assessment Approaches

Regarding quantitative impact assessment approaches, the main emphasis is on quantitatively assessing the degree and extent of impacts. To evaluate the success of an intervention and determine any adverse effects, some level of quantification is often necessary for impact assessments. These approaches typically rely on micro-economic approaches that follow econometric models.

There are specific characteristics associated with quantitative impact assessment approaches. Firstly, they conduct baseline studies to identify the initial conditions, define objectives, set targets, and rigorously evaluate performance and outcomes. However, it is essential to note that these approaches can be costly and have limited scope. They may only accurately measure specific types of impacts and inferring cause-and-effect relationships can pose challenges.

#### *7.4.6.1. Types of Quantitative Impact Assessment Approaches*

##### *7.4.6.1.1. Based on the Period of Assessment*

Impact assessment may occur before approval of an intervention (ex-ante), after completion (ex-post), or at any stage.

- *Ex-ante Assessment*
  Ante assessment is crucial to the policy cycle's initial stages, focusing on analyzing and planning potential impacts. It involves a forward-looking analysis of an intervention's effects to guide decision-making, akin to strategic planning in the business world. Ante assessment predicts the possible outcomes as part of the intervention's planning, development, and approval process. An illustration of this process is the evaluation of the introduction of a Genetically Modified Rice variety.
- *Ex-post Assessment*
  It is an integral part of the evaluation and management phase within the policy cycle. Evaluation, in general, seeks to determine the extent to which a policy intervention addresses the targeted issue and how it does so. Impact assessment, on the other hand, specifically focuses on the outcomes of the intervention. Evaluation typically encompasses a broader array of considerations, including the effectiveness of the intervention's design, its cost-efficiency, unintended consequences, and how lessons learned can enhance future intervention strategies. Post-assessment identifies the actual impacts during and after implementation, facilitating necessary corrective measures and providing insights for improving future intervention designs. For instance, it could involve assessing the knowledge acquisition of trainees from a multimedia compact disc (CD) on the System of Rice Intensification (S.R.I.) method of rice cultivation.

*7.4.6.1.2. Based on the Research Design*

Impact assessment may be conducted using experimental, quasi-experimental, and non-experimental designs.

*A. Experimental Approaches*

- *Randomized Evaluations*
  Randomized evaluations involve extending the benefits of a program to a treatment group that is randomly selected while maintaining an identical control group. The progress of both the treatment and control groups is monitored over time, specifically focusing on selected impact indicators (Khandker et al., 2010). Randomized evaluations

are employed when the eligible population is sufficiently large to implement the program. The program is gradually implemented until it encompasses the entire eligible population. One notable advantage of this approach is its capacity to eliminate bias in selecting respondents.

- *Pre-Test/Post-Test with Random Assignment to Intervention or Comparison Groups*
  In randomized experiments, participants are randomly assigned to either a study group that receives the technological intervention or a comparison group that does not. Data is collected before and after the intervention for each group. The differences observed between the intervention and comparison groups at the end of the experiment can be directly attributed to the impact of the intervention, provided that the sample size is sufficiently large. These methods can also be applied to small sample sizes of less than 30 individuals per group (Pitman, 1937). The selection of statistical techniques is based on the specific research question at hand, and repeated measures ANOVA is utilized to determine whether there is a significant difference in the mean change of outcomes before and after treatment between the two groups. On the other hand, to assess whether the post-test means, adjusted for pre-test scores, differ between the two groups, Analysis of Covariance (ANCOVA) is employed.

Two groups are randomly assigned the subjects and treatment conditions. Data are collected only after the intervention. Statistical tests are regression, t-test, and ANOVA.

*B. Quasi-Experimental Approaches*

*(i) Pre-Test/Post-Test with Non-Random Assignment to Intervention or Control Groups*

In this research design, data is gathered both before and after the intervention, with subjects being assigned to either the intervention or control groups in a non-random manner. The comparison groups in quasi-experimental designs can be distinguished through propensity score matching (McKienzie & Cusolito, 2017).

- *Propensity Score Matching –*
  The propensity score matching method has gained popularity for estimating the causal treatment effects (Caliendo & Kopeninig, 2008; Rubin & Thomas, 2000). It involves selecting a control population (non-beneficiaries) by matching them with the actual beneficiaries based on observable characteristics such as personal and socioeconomic attributes (Diaz & Honda, 2006). The matched control groups can be chosen either before the project implementation (prospective studies) or after (retrospective studies) (Perkins et al., 2000). Various approaches are used for matching beneficiaries and non-beneficiaries, including nearest-neighbor (N.N.) matching, caliper and radius matching, stratification and interval matching, kernel matching, and local linear matching (L.L.M.) (Khandker et al., 2010; Raudenbush & Bryk, 2002).
- *Statistical analysis – Reliability-corrected ANCOVA.*
  LaLonde (1986) utilized this method to assess the impact of a training program on trainee earnings, while Dehejia and Wahba (1999) evaluated the impact of a training program using a matching estimator based on National Supported Work (N.S.W.) data.

*(ii) Two-Group Post-Test Only with Non-Random Assignment*
Data is collected solely after the program has concluded from participants who received the intervention and non-participants. Matching participants and non-participants based on similar characteristics and addressing any relevant differences is crucial in the post-test-only design to isolate the effects of the intervention.

*(iii) Double Difference or Difference-In-Differences (D.I.D.) Methods*
Estimating the impact of a specific intervention involves comparing changes in outcomes over time between a group enrolled in the program (intervention group) and a group not enrolled (control group). This method can be utilized in experimental and quasi-experimental designs, necessitating baseline and follow-up data from the treatment and control groups. Before the intervention, a baseline survey is conducted to evaluate outcome indicators among beneficiaries (treatment group) and non-beneficiaries (control or comparison group), allowing for a comparison to determine differences (Khandker et al., 2010). After the intervention, a post-intervention survey assesses differences between the treatment and control groups. The mean difference between the

post-intervention and pre-intervention values of outcome indicators for each group is calculated, followed by the difference between these mean differences. This second difference, known as the difference in differences, estimates the program's impact. Statistical methods such as Repeated Measures ANOVA, Repeated Measures Linear Regression Analysis, and Propensity Score Matching are commonly used when randomization at the individual level is not feasible (Lee et al., 2010).

*(iv) Drawbacks of the Difference-in-Differences Method*

Even if trends are initially the same before the implementation of an intervention, there is a possibility of bias in the difference-in-differences estimation that may remain undetected. This is because the difference-in-differences method attributes any variations in trends between the group receiving the treatment and the comparison group to the time when the intervention is introduced (Limieux & Milligun, 2005). The estimation could be flawed or biased if other factors influence other differences in trends between the two groups that are not considered in the multivariate regression analysis. For instance, when attempting to assess the impact of subsidizing fertilizers on rice production by comparing the production of rice from farmers who received subsidies (treatment group) and those who did not (comparison group) before and after the subsidy distribution, any external factors affecting the production levels could lead to inaccurate results. If, for example, there is a drought that only affects the subsidized farmers in the first year, the difference-in-differences estimate would provide an incorrect assessment of the impact of fertilizer subsidies. In essence, any factor disproportionately affects one group over the other while the treatment is administered. It is not considered that the regression analysis has the potential to distort or bias the estimation of the program's impact. The difference-in-differences method assumes the absence of such factors.

*(v) Regression Discontinuity Design*

The regression discontinuity design method is commonly used in impact evaluation to determine the eligibility of participants for a program with a continuous eligibility index and a clearly defined cut-off score. This method involves assigning participants to intervention or comparison groups based on whether they fall above or below the cut-off score on a pre-intervention measure (Burton et al., 1998). The average treatment effect can be estimated by comparing observations closely on either side of the threshold. This design is particularly useful when randomization is not feasible, and the researcher wants to target the intervention to those who are most in need or deserving of it. There are two types of regression discontinuity designs. The Sharp method assigns a discrete cut-off point for both beneficiaries and non-beneficiaries and compares the means of treatment effects to assess the impact. On the other hand, the fuzzy method is used when a few eligible individuals are excluded from the intervention or become ineligible for different reasons. The fuzzy version of the technique is more widely used.

Duflo (2003) employed this method to study the impact of an expanded old-age pension program by using the eligible age requirement for men and women as instruments for receiving a pension. Other studies, such as the one by Buddlemeyer and Skofias (2003), have also presented regression discontinuity as an alternative to experimental methods. Cook, Shadish, and Wong (2006) supported the findings of Buddlemeyer and Skofias (2003) by stating that randomized and nonrandomized experiments yield similar results when the regression discontinuity method is applied.

Various studies have utilized the RDD technique, but it has limitations. These include the presence of specification errors and the necessity for large evaluation samples to ensure adequate statistical power when employing RDD. Despite these challenges, RDD provides an unbiased estimation of the impact of the eligibility cut-off in the local area.

The eligibility index must exhibit continuity around the cut-off score, and units should demonstrate similarity in the vicinity above and below this threshold.

- It is crucial to assess whether there is any noncompliance with the rules dictating treatment eligibility. This can be done by verifying that all eligible units have received treatment while no ineligible units have.
- It is essential to investigate whether there is any indication of manipulation in the index scores to influence program qualification. Testing for a smooth distribution of index scores at the cut-off point can reveal any potential manipulation, such as clustering scores above or below the threshold.
- Lastly, it is essential to determine whether the cut-off is specific to the program under evaluation or if it is utilized by other programs as well.

*(vi) Limitations of the Regression Discontinuity Design Method*

Estimating local average treatment effects (LATE) using regression discontinuity design focuses on the eligibility cut-off point, where treatment and comparison units exhibit the highest degree of similarity. As one approaches the cut-off point, the units on both sides become increasingly similar. When one reaches a score very close to the cut-off, the units on either side become so alike that the comparison becomes as reliable as if the treatment and comparison groups had been randomly assigned. Since the RDD method estimates the program's impact specifically around the cut-off point or locally, it is essential to note that the estimate may not apply to units whose scores are further away from the cut-off. In such cases, eligible and ineligible individuals may differ, making the forecast less generalizable. The fact that the RDD method cannot provide an average treatment effect for all program participants can be viewed as both a strength and a limitation, depending on the evaluation question. Suppose the primary objective is to determine whether the program should exist. In that case, the average treatment effect for the eligible population becomes the most relevant parameter, and the RDD method to be refined accordingly. However, if the policy question of interest pertains to whether the program should be scaled back or expanded at the margin, specifically for (potential) beneficiaries near the cut-off point, then the RDD method precisely generates the local estimate that is crucial for informing this significant policy decision. Since the RDD method estimates the impact of the program specifically around the cut-off point or locally, it is essential to note that the estimate may not be applicable to units whose scores are further away from the cut-off. In such cases, eligible and ineligible individuals may differ, making the estimate less generalizable. The fact that the RDD method cannot provide an average treatment effect for all program participants can be viewed as both a strength and a limitation, depending on the evaluation question at hand. If the primary objective is to determine whether the program should exist, then the average treatment effect for the eligible population becomes the most relevant parameter, and the RDD method must be refined accordingly. However, if the policy question of interest pertains to whether the program should be scaled back or expanded at the margin, specifically for (potential) beneficiaries in close proximity to the cut-off point, then the RDD method precisely generates the local estimate that is crucial for informing this significant policy decision.

*C. Non-Experimental Designs (MLE, 2013)*

These designs consist solely of an intervention group, lacking a control group, making them the least robust designs for assessing impact. They are typically employed in situations (i) limited resource conditions, (ii) when researchers are unable to create a comparison group, and (iii) when an intervention covers the entire population.

1. *Pre-Test/Post-Test Designs*
   In pre-test/post-test designs, the researcher measures changes in the specific phenomenon (outcome indicator) among subjects before and after the intervention. However, it is essential to note that this design alone cannot attribute all the changes to the intervention, as there is no comparison group.
2. *Time-Series Designs*
   Time-series designs involve estimating changes in outcome indicators over time to determine trends. Data is collected multiple times before and after the intervention to these trends.
3. *Longitudinal Study*
   In a longitudinal study, the researcher records repeated measures of the same variables from the same individuals. A panel design, a particular type of longitudinal design, involves tracking a smaller group of people at multiple points and recording their experiences in detail.
4. *Post-Test Only Design*
   In an only post-test-only design, researchers observe the intervention group at one point after the intervention. The focus is mainly on comparing sub-group responses based on characteristics such as age, sex, ethnicity, education, or level of exposure to the intervention.

*Methods to Strengthen Non-Experimental Impact Assessment*

a. Measure participants' exposure to the program: By measuring the participants' initial exposure to the intervention, selection bias can be offset.
b. Collect data from the same participants over time using a panel or longitudinal design: This allows individuals to serve as their control.

*D. Instrumental Variable (IV) Methods*

The instrumental variable (IV) is a statistical estimation method used in non-experimental impact analysis designs. Including a new variable in the analysis helps minimize unobservable biases in sample selection. This new variable, the instrumental variable, increases the probability of being selected as a beneficiary in the study but does not influence the intervention outcome. For example, 'geographical variation' may be included as an instrumental variable when assessing the impact of the poverty alleviation program (Ravallion 2005).

*Randomized Promotion as an Instrumental Variable*

Randomized promotion ensures that the counterfactual estimates are valid when the promotion campaign significantly boosts program participation without directly influencing the outcomes. Are the baseline characteristics evenly distributed between the units exposed to the promotion campaign and those not? Contrast the baseline characteristics of the two groups. Does the promotion campaign substantially impact the program's participation rate? It is expected to. Compare the participation rates in the program between the promoted and non-promoted subgroups. Does the promotion campaign have a direct impact on the outcomes? It is not supposed to. This cannot be directly examined; one must depend on theory, common sense, and a thorough understanding of the impact evaluation context for guidance. Randomized promotion ensures that the counterfactual estimates are valid when the promotion campaign significantly boosts the participation in the program without influencing the outcomes directly. Are the baseline characteristics evenly distributed between the units exposed to the promotion campaign and those not? Contrast the baseline characteristics of the two groups. Does the promotion campaign substantially impact the program's participation rate? It is expected to. Compare the participation rates in the program between the promoted and non-promoted subgroups. Does the promotion campaign have a direct impact on the outcomes? It is not supposed to. This cannot be directly examined; one must depend on theory, common sense, and a thorough understanding of the impact evaluation context for guidance.

Towards the end of the discussion on impact evaluation approaches, Table 7.1 provides a comprehensive comparison of different methodologies for impact assessment. This table offers a detailed overview of various approaches, including their descriptions, comparison group compositions, key assumptions, and required data types. As shown in Table 7.1, methodologies range from Randomized Assignment to Instrumental Variable methods, Regression Discontinuity Design, Difference-in-Differences, and Matching approaches, each with unique characteristics and research applications.

**Table 7.1.** Comparison of different approaches to impact evaluation

| Methodology | Description | Who is in the Comparison Group? | Key Assumption | Required Data |
|---|---|---|---|---|
| Randomized Assignment | Eligible units are assigned at random to treatment or comparison groups. Every eligible unit has an equal probability of being selected. Internally valid effect estimates are usually produced, even under the weakest assumptions. | Eligible units that are randomly assigned to the comparison group. | Randomization effectively produces two groups that are statistically identical to observed and unobserved characteristics (at baseline and through the online). | Follow-up outcome data for both treatment and comparison groups; baseline outcomes and other characteristics for both groups to ensure balance. |
| Instrumental Variable (particularly randomized promotion) | A randomized instrument, such as a promotion campaign, influences participation in the evaluated program. This method uses the changes in outcomes resulting from the changes in participation rates to estimate program impacts. | "Compiler" units whose participation in the program is influenced by the instrument (they would participate if exposed to the instrument, but not if not exposed) | The instrument affects program participation but does not directly affect outcomes (i.e., the instrument only affects outcomes by altering the probability of participating in the program) | Follow-up outcome data for all units; data on effective participation in the program; data on baseline outcomes and other characteristics. |
| Regression Discontinuity Design | Units are ranked based on specific quantitative and continuous criteria, such as a poverty index. A cut-off determines whether a unit is eligible to participate in a program. Outcomes for participants on one side of the cut-off are compared with outcomes for non-participants on the other. | Units close to the cut-off but ineligible to receive the program. | Units immediately below and above the cut-off are statistically identical, allowing us to identify unbiased program impacts for the population close to the cut-off. | Follow-up outcome data; ranking index and eligibility cut-off; data on baseline outcomes and other characteristics. |

| Methodology | Description | Who is in the Comparison Group? | Key Assumption | Required Data |
|---|---|---|---|---|
| Difference-Indifferences | The change in outcome over time in a group of non-participants is used to estimate what would have been the change of outcomes for a group of participants in the absence of a program. | Units that did not participate in the program (for any reason) and for which data were collected before and after the program. | If the program had not existed, the outcomes for the participants and non-participants would have grown in parallel over time. | Baseline and follow-up data on outcomes and other characteristics for participants and non-participants. |
| Matching (particularly propensity score matching) | For each program participant, the method looks for the "most similar" unit in the non-participant group (the closest match based on observed characteristics). | Based on observed characteristics, the non-participant unit is predicted to have the same likelihood of participating in the program as each participant. | No characteristic affects program participation beyond the observed characteristics used for matching. | Follow-up outcome data for participants and non-participants; data on effective participation in the program; baseline characteristics to perform matching. |

## 7.5. Economic Impact Assessment

Economic Impact Assessment (E.I.A.) is a process used to evaluate the financial consequences of a particular intervention, project, policy, or event. It aims to understand how the intervention affects various aspects of the economy, including employment, income, output, productivity, investment, and fiscal revenues.

### 7.5.1. Partial Budgeting Technique (PBT)

Partial Budgeting Technique (PBT) is a financial analysis method used in agricultural and business management to evaluate the economic impact of specific changes or decisions. It assesses the costs and benefits associated with a particular management practice, technology adoption, investment, or operational change within a defined timeframe. PBT helps decision-makers weigh the financial implications of alternative options and determine whether a proposed change is economically viable or profitable. An illustration of this approach can be seen in examining the impact of implementing drip irrigation in tomato cultivation. This method encompasses four key elements:

- Augmentation of revenue;
- Diminishment or eradication of expenses;
- Escalation of expenses;
- Diminishment or eradication of revenue.

$$\text{Profit/Loss} = [\text{Added returns} - \text{Added costs}] + [\text{Reduced costs} - \text{Reduced returns}]$$

Characteristics of this approach:

- The method is straightforward and can be easily grasped;
- It focuses solely on the net changes in costs and benefits, making it suitable for evaluating the economic feasibility of individual intervention technologies;
- It demands less data compared to whole-farm budgeting as it does not consider fixed costs;

- It enables prompt assessments regarding the suitability of the new technology.

### 7.5.2. Net Present Value (N.P.V.)

The Net Present Value (N.P.V.) serves as a straightforward metric to determine the feasibility of an investment. It quantifies the amount of cash that an investor would require at present in lieu of investing. A positive N.P.V. indicates that the investment is advantageous, as it is akin to receiving a cash sum equivalent to the N.P.V. Conversely, a negative N.P.V. suggests that investing today would result in a loss of cash, warranting rejection of the investment. When the anticipated return matches the discount rate, N.P.V. equals zero, rendering the investor indifferent toward the investment. N.P.V. is calculated as the present value of all benefits discounted at the appropriate rate, subtracted by the present value of all costs discounted at the same rate. Symbolically, it can be expressed that:

$$NPV = \sum_{0}^{t} \frac{(B_t - C_t)}{(1 + r)^t}$$

where: $B_t$ is the benefit at time t
$C_t$ is the cost at the time t
r is the discount rate
t refers to the time

### 7.5.3. Benefit-Cost Ratio (B.C.R.)

The benefit-cost ratio is a method that transforms the time series data of benefits and costs, which are utilized in the calculation of Net Present Value (N.P.V.), into a ratio format instead of presenting them as individual values. Another way to describe the B.C.R. is by comparing an investment's discounted benefits and costs at a specific time.

$$BCR = \frac{\sum B_t(1 + r)^t}{\sum C_t(1 + r)^t}$$

### 7.5.4. Internal Rate of Return (IRR)

The internal rate of return (IRR) is a metric that indicates the rate at which the capital invested in a business grows. When analysing the economic effects of research and development (R&D), IRR is referred to as the private rate of return (P.R.R.) when studying the return on investment for a single company's R&D efforts or as the social rate of return (S.R.R.) when estimating rates of return for an entire industry or economy. IRR is the rate at which the net present value (N.P.V.) equals zero, and it can be easily calculated using the GOAL SEEK function in the WHAT-IF ANALYSIS tool of MS EXCEL. Alternatively, the formula provided below can also be used.

$$IRR = r_1 + \frac{NPV_1}{NPV_1 - NPV_2}(r_2 - r_1)$$

$r_1$ = lower discount rate chosen NPV1= NPV at $r_1$
$r_2$ = higher discount rate chosen NPV2= NPV at $r_2$

#### *7.5.4.1. Cost Concepts*

The following cost concepts can be used to measure the profitability of new technology.

"Cost $A_1$ = All actual expenses in cash and kind incurred in production."
"Cost $A_2$ = Cost $A_1$ + Rent paid for leased in the land."
"Cost $B_1$ = Cost $A_1$ + Interest on value of owned capital assets"
"Cost $B_2$ = Cost $B_1$ + Rental value of owned land and rent paid for leased land"
"Cost $C_1$ = Cost $B_1$ + Imputed value of family labour"
"Cost $C_2$ = Cost $B_2$ + Imputed value of family labour"
"Cost $C_3$ = Cost $C_2$ + 10% of Cost $C_2$ on account of managerial functions performed by the farmer"

Other significant instruments include:

- Utilization rate/indicator/ratio
- Variables that impact the acceptance of technology (logit/probit)
- Model for consumer surplus
- Regression models with dummy variables

- Alteration in time series models
- Model for input-output/S.A.M.
- Index for competitiveness (for exports)

## 7.6. Choosing Whether to Conduct an Impact Assessment

Not all programs necessitate an impact evaluation process. The Impact evaluations should be employed selectively when there is a need for a comprehensive examination of causality. They can be expensive if one must collect data, and it is important to allocate the evaluation budget strategically. When considering the expansion of a new program and contemplating whether to proceed with an impact evaluation, it is beneficial to ask a few fundamental questions.

The initial question to consider is the significance of the matter at hand. Will evidence of the program's success, modality, or design innovation contribute to crucial decisions? These decisions often involve budget allocations and program scalability. If the budget implications are limited or the results will only affect a small number of individuals, conducting an impact evaluation may not be worthwhile.

If it is determined that the stakes are high, the next question to consider is whether there is existing evidence to demonstrate the program's effectiveness. Specifically, does one know the magnitude of the program's impact? Is there available evidence from similar programs in similar circumstances? If there is no evidence regarding the potential of the program being considered, it is advisable to begin with a pilot program incorporating an impact evaluation. Conversely, if evidence is available from similar circumstances, the cost of an impact evaluation will likely be justified only if it can address an essential and novel policy question. This would be the case if the program included significant innovations that still need to be tested.

To warrant the allocation of technical and financial resources required for conducting a high-quality impact assessment, the intervention under evaluation should possess the following characteristics:

- *Innovative.* It should explore a new, promising approach.
- *Replicable.* It should be scalable or adaptable to different settings.

- *Strategically relevant.* The findings from the impact assessment should influence a crucial decision regarding the intervention, such as program expansion, reform, or budget allocations.
- *Untested.* Limited information on the program's effectiveness or alternative designs is available globally or in a specific context.
- *Influential.* The results will be utilized to shape policy decisions.

Do we possess the essential resources for a comprehensive impact evaluation? These resources encompass technical aspects like relevant data and time, financial support of the assessment, and institutional support for the involved teams, along with their dedication to generating and utilizing causal evidence. An evaluation team comprises policymakers and researchers collaborating to properly implement a well-designed, technically sound evaluation that yields pertinent results for critical policy and program design inquiries. A thorough grasp of the purpose and potential of impact evaluation by the evaluation team is crucial for its success.

## 7.7. Ethical Considerations Regarding Impact Evaluation

When developing an impact evaluation, it is crucial to consider various ethical considerations. There have been discussions regarding the moral implications of conducting impact evaluations. One key aspect of this debate involves examining the ethics of allocating significant public resources to programs whose effectiveness is still unknown (Karlan & Zinman, 2010). In this light, the absence of evaluation may be viewed as unethical. The data on program effectiveness produced by impact evaluations can contribute to a more efficient and ethical allocation of public resources.

Furthermore, ethical concerns also revolve around the criteria used to distribute program benefits, the treatment of human subjects during studies, and the transparency in documenting research plans, data, and outcomes. A fundamental ethical principle in evaluations is that interventions with proven benefits should not be withheld or delayed solely for the sake of evaluation. Evaluations should not determine how benefits are distributed; instead, they should align with program allocation criteria that are fair and transparent. Any ethical issues related to program allocation rules do not arise from the impact evaluation itself but from the operational guidelines of the program. Planning evaluations can help clarify program guidelines and assess their fairness and transparency based on specific eligibility criteria (Gertler et al. 2008; Gertler et al., 2007).

The random assignment of program benefits often raises ethical concerns about denying benefits to eligible participants. However, many programs operate within limited financial and administrative resources, making it impractical to reach all eligible participants simultaneously. From an ethical perspective, all individuals equally eligible for a social program should have an equal chance of receiving benefits. Random assignment meets this moral standard.

When implementing a program gradually, introducing it can involve randomly selecting the order in which equally deserving beneficiaries will receive the program. In such cases, beneficiaries who join the program later can serve as a comparison group for those who entered earlier. This approach allows for a robust evaluation design and a fair method of allocating limited resources. The ethical considerations in impact evaluation extend beyond the rules governing program assignment. They also encompass the ethical treatment of human subjects in research and the principles of conducting transparent, objective, and reproducible studies.

Review boards or ethics committees have been established in numerous countries and international institutions to ensure the ethical conduct of research involving human subjects to ensure the ethical conduct of research involving human subjects. These boards are responsible for evaluating, approving, and monitoring research studies, with the primary aim of safeguarding the rights and well-being of all participants. While impact evaluations primarily involve operational activities, they are also considered research studies and should adhere to guidelines for research involving human subjects. A crucial ethical aspect of research is to make impact evaluations objective, transparent, and reproducible.

To promote transparency in research, impact evaluation plans can be included in a pre-analysis plan and submitted to a study registry. Furthermore, once the research is completed, the data and code used in the analysis should be publicly available. This allows others to replicate the study while ensuring anonymity and protection.

## 7.8. Challenges in Extension Programs Evaluation and Impact

Results of extension programs particularly change in Knowledge, Attitudes, Skills, and Aspirations (KASA) and long-term outcomes, are often intangible and difficult to measure. Suvedi and Vander Stoep (2014) identified three main challenges in evaluating extension programs:

- The frequent use of descriptive, one-shot case studies for impact evaluation highlights the need for precise assessments of cause-and-effect outcomes. Extension programs must establish a scientific knowledge base and generate evidence. For example, a key question is whether farmers who participate in extension training adopt new technologies sooner than those who do not.
- Difficulty in establishing a control or comparison group, maintaining consistent treatments, and measuring the long-term impacts of extension programs.
- Lack of baseline and monitoring data on impact indicators hinders the ability to track expanded impacts over time.

Evaluating extension programs is essential to ensure their effectiveness and impact on the intended beneficiaries. However, various challenges, from methodological limitations to resource constraints, complicate the evaluation process. This paper delves into these challenges and proposes strategies to overcome them, ultimately improving the evaluation of extension programs and the impact assessment of one major obstacle: setting clear and measurable objectives. These programs often have broad goals, making establishing specific and attainable objectives challenging. Moreover, objectives may differ among stakeholders, further complicating the evaluation process. Addressing this challenge necessitates stakeholder collaboration to define objectives using the SMART criteria (Specific, Measurable, Achievable, Relevant, Time-bound).

Another obstacle involves the selection of appropriate evaluation methodologies. Extension programs function in various settings, necessitating adaptable and contextually relevant evaluation approaches. However, identifying suitable methods that can effectively capture the complexity of program outcomes while remaining practical and feasible can be time-consuming and demanding. By integrating quantitative and qualitative data collection techniques, mixed-method approaches can address this challenge by offering a comprehensive understanding of program impacts.

Resource limitations present a significant hurdle to the practical evaluation of extension programs. Scarce funding, personnel, and time can impede thorough evaluation activities. To surmount this obstacle, using existing resources, such as collaborating with academic institutions or engaging community volunteers, can help supplement evaluation endeavors. Moreover, prioritizing evaluation activities based on their potential for learning and improvement can ensure the efficient utilization of limited resources.

Ensuring stakeholder engagement and buy-in poses another challenge in extension program evaluation. Stakeholders may possess divergent interests and priorities, resulting in resistance or lacking support for evaluation initiatives. Establishing trust, promoting open communication, and involving stakeholders throughout the evaluation process can aid in mitigating this challenge. Engaging stakeholders in defining evaluation objectives, gathering data, and interpreting findings fosters ownership and enhances the likelihood of utilizing evaluation results for decision-making.

Additionally, measuring long-term impacts and sustainability presents challenges in extension program evaluation. Extension interventions often strive to instigate enduring changes in behaviors, practices, and systems, which may require time to take root. Monitoring and measuring these long-term impacts necessitate longitudinal studies and continuous monitoring and evaluation efforts. Furthermore, evaluating the sustainability of program outcomes entails considering factors such as community capacity building and policy support.

## 7.9. Core Elements of a Well-Designed Impact Evaluation

Impact evaluations involve intricate processes with various components. The subsequent checklist outlines the fundamental aspects of a well-structured impact evaluation (NCVO, 2012)

- A specific and pertinent policy query—based on a theory of change—that can be addressed through an impact evaluation
- A strong methodology, originating from the operational guidelines of the program, to calculate a counterfactual demonstrating the causal connection between the program and desired outcomes
- An adept evaluation team that operates as a collaboration between a policy team and a research team
- Adherence to ethical norms and consideration of human subjects in the planning and execution of the evaluation and associated data

collection, along with a focus on open science principles for transparency
- A sample size with adequate statistical power to identify policy-relevant impacts
- A methodology and sample that yield results applicable to the target population
- High-quality data supplying the necessary information for the impact evaluation, encompassing data for the treatment and control groups, baseline and follow-up data, and details on program implementation and expenses
- An engagement plan to facilitate policy discussions through the implementation of the impact evaluation, as well as prompt disseminating an impact evaluation report and relevant policy briefs to key stakeholders.

## 7.10. Tips to Mitigate Common Risks in Conducting an Impact Evaluation

We also emphasize specific recommendations that can assist in mitigating common risks inherent in the process of conducting an impact evaluation:

- Impact evaluations are most effectively planned early in the project cycle, preferably as part of the program design or at least before the evaluation program is implemented. Initiating the planning phase early on allows for the development of a prospective evaluation design based on the most reliable methodology available (Levy & Ohls, 2007). This approach provides sufficient time to strategize and execute the collection of baseline data in the evaluation areas before the commencement of the program.
- The results of impact evaluations should be informed by complementary process evaluations and monitoring data demonstrating the program's implementation. Understanding the reasons behind the success or failure of programs is crucial. It is essential to differentiate between a poorly executed program and a flawed program design. Collecting baseline data and incorporating a backup methodology into the impact evaluation design is vital. If the original evaluation design becomes invalidated, such as when the

original comparison group receives program benefits, having a backup plan in place can prevent the need to discard the entire evaluation.

- Maintaining standard identifiers among data sources is imperative to facilitate the easy linkage of observation units during the analysis. For instance, a specific household should have the same identifier in the monitoring systems, baseline data, and follow-up impact evaluation surveys.
- Impact evaluations serve as valuable tools for understanding how programs operate and for testing program alternatives, even in the case of large ongoing programs. Well-designed impact evaluations can help assess innovations or provide insights into the relative effectiveness of different goods and services delivered as part of existing programs. By incorporating an additional program innovation as a small pilot within the context of a more extensive evaluation, it is possible to leverage the evaluation process to generate valuable information for future decision-making.
- This should be considered an integral component of a program's operation and should be adequately staffed.

## Conclusion

Evaluating the impact and measuring the success of extension activities are essential steps in ensuring their effectiveness and sustainability. By employing systematic evaluation processes, organizations can assess how their programs achieve their intended objectives and make a difference in the lives of individuals and communities. Clear goals and outcomes provide a roadmap for extension activities, guiding the selection of appropriate strategies and interventions. Key performance indicators (KPIs) enable organizations to quantify progress and success, helping them track performance and identify areas for improvement. Baseline assessments establish a starting point against which progress can be measured, providing valuable insights into the context and challenges faced by target populations.

Monitoring and data collection allow organizations to track the implementation of activities in real time, ensuring that interventions are delivered as planned and reach the intended audience. Regular monitoring also enables organizations to identify emerging issues and make timely adjustments to program delivery. Evaluation of impact provides an opportunity to assess the effectiveness of extension activities, identifying successes and areas for improvement. By analyzing quantitative and qualitative data, organizations can gain a comprehensive understanding of the outcomes and impacts of their programs. Feedback and reflection play a critical role in the evaluation process, allowing organizations to learn from their experiences and make informed decisions about program design and implementation. Long-term evaluation provides insights into the sustained impact of extension activities over time, helping organizations understand the lasting effects of their interventions on individuals, communities, and systems.

Stakeholder engagement is integral to the evaluation process, ensuring that the perspectives and priorities of key-stakeholders are considered throughout program design, implementation, and evaluation. By involving stakeholders in the evaluation process, organizations can build trust, foster collaboration, and enhance the relevance and effectiveness of their extension activities. Evaluating the impact and measuring the success of extension activities are essential for maximizing their effectiveness and ensuring that resources are used efficiently. By employing systematic evaluation processes, organizations can assess the outcomes and impacts of their programs, identify areas for improvement, and make evidence-based decisions about future programming. Ultimately, effective evaluation contributes to achieving sustainable development goals and empowering individuals and communities to build a better future.

## References

Angrist, J, Bettinger, E., Bloom, E., King, E., & Kremer, M., (2002). Vouchers for private schooling in Colombia: Evidence from a Randomised Natural Experiment, *American Economic Review*, 92(5), 1535-38.

Angrist, J., & Krueger, A. B. (2001). Instrumental variables and the search for identification: From Supply and Demand to Natural Experiments, *NBER*, Working Paper No. 8456.

Angrist, J., & Lavy, V. (1999). Using Maimonides to estimate the effect of class size on scholastic achievement, *Quarterly Journal of Economics*, 114 (2), 533-575.

Badioli, F. (2011.) Indicators of Impact. Paper presented at the Training Workshop on Sustainability & Embedding. Euromed Heritage.

Baker, J. L. (2000.) Evaluating the impact of development projects on poverty: *A handbook for practitioners.* Washington D.C.: World Bank.

Blankenburg, F. (1995). *Methods of Impact Assessment Research Programs, Resource Pack and Discussion.* The Hague: Oxfam UK/I and Novib.

Buddlemeyer, H., & Skofias, E. (2003). An Evaluation on the Performance of Regression Discontinuity Design on PROGRESA, *Institute for Study of Labor*, Discussion Paper No.827.

Burton, P., Gurrin, L., & Sly, P. (1998). Extending the simple linear regression model to account for correlated responses: An introduction to generalized estimating equations and multi-level mixed modelling. *Statistics in Medicine, 17*(11), 1261–1291. https://doi.org/10.1002/(sici)1097-0258(19980615)17:11<1261::aid sim846> 3.0.co;2-z.

Burton, P., Gurrin, L., & Sly, P. (1998). Tutorial in biostatistics, extending the simple linear regression model to account for correlated responses: An introduction to the Generalized estimating equations and multi-level modeling, *Statistics in Medicine*, 17, 1261-1291.

Caliendo and Kopeinig (2008). Some practical guidance for the implementation of propensity score matching, *Journal of Economic Survey*, 22(1), 31-72.

Cook, T. D., Shadish, W. R., & Wang, V. C. (2006). Three conditions under which experiments, and observational studies produce comparable causal estimates: New findings from within-study comparisons, *Journal of Policy Analysis and Management*, 27(4), 724-750.

Dart, J., Petheram, R. J. and Staw, W (1998). *Review of Evaluation in Agricultural Extension.* Rural Industries Research and Development Corporation, Australia, 107p.

Dehejia, R. (2005). Practical Propensity Score Matching: A reply to Smith and Todd, *Journal of Econometrics*, Vol.125 (1-2), 355-364.

Dehejia, R. H., & Wahba, S. (2002). Propensity scores matching methods for non-experimental causal studies, *Review of Economics and Statistics*, 84(1), 151-161.

Dehejia, R., & Wadba, S. (1999). Causal effects in non-experimental studies: Re-evaluating the evaluation of training programs, *Journal of the American Statistical Association*, 94, 1053-1062.

Diaz, J. J., & Honda, S. (2006). An assessment of Propensity Score matching as a non-experimental impact estimator: Evidence from a Mexican Poverty Program, *The Journal of Human Resources*, 41(2), 319-345.

Duflo, E. (2001). Schooling and labour market consequences of school construction in Indonesia: Evidence from an unusual policy experiment, *American Economic Review*, 9(4), 795-813.

Duflo, E. (2003). Grandmothers and granddaughters: Old age pension and intra household allocation in South Africa, *World Bank Economic Review,* 17(1), 1-26.

Duflo, E., & Pande, R. (2007). Dams, *Quarterly Journal of Economics,* 122 (2), 601–646.

Filmer, D., & Schady, N. (2009). School Enrollment, Selection and Test Scores, *World Bank Policy Research* Working Paper 4998, World Bank, Washington DC.

Gertler, P., Martinez. S., & Vivo, S. (2008). Child-mother provincial investment project plan Nacer, University of California Berkley and World Bank, Washington DC.

Gertler, P., Martinez, S., Premand, P., Rawling, B. L., & Vermeersch. J. M. C., (2007). Impact evaluation in practice, *Washington, DC: World Bank.*

Gertler, P., Martinez, S., Premand, P., Rawling, B. L., & Vermeersch. J. M. C., (2009). Impact evaluation in practice, *Washington, DC: World Bank.*

Heckman, J., Ichimura, H., & Todd, P. E., (1998). Characterizing selection bias using experimental data, *Econometrica,* 66(5),), ￼7-1098.

Imas, L. G. M., & Rist, R. C. (2009). The Road to Results: Designing and conducting Effective Development Evaluations, *Washington, DC: World Bank.*

Johnson, B., & Onwuegbuzie, A. (2004). Mixed Methods Research: A Research Paradigm Whose Time Has Come, *Educational Researcher,* 33(7), 14-26.

Karlan, D., & Zinman, J. (2010). Expanding credit access using randomized supply decisions to estimate the impacts, *Review of Financial StudStudies,* ￼ety for Financial Studies, 23(1), 433-464.

Khandker, R. S., Koolwal, G. B. and Samad, S. A. (2010.) Handbook on Impact Evaluation: Quantitative Methods and Practices. Washington, D.C.: World Bank.

Lalonde, R. J. (1986). Evaluating the econometric evaluations of training programs with experimental data, *The American Economic Review*, 76(4), 604-620.

Lee, B. K., Lessler, J., & Stuart, E. A. (2010). Improving propensity score weighting using machine learning, *Statistics in medicine*, 29,337-346.

Levy, D., & Ohls, J. (2007). Evaluation of Jamaica's PATH Program: Final Report, *Mathematica Policy Research, Inc*; Ref. 8966-090, 090, ￼ngton DC.

Limieux, T., & Milligun, K. (2005). Incentive Effects of Social Assistance: A Regression Discontinuity approach, NBER Working Paper 10541, *National Bureau of Economic Research,* Cambridge, MA.

McKienzie, D., Assaf, N., & Cusolito, A. P. (2017). The additionality impact of a matching grant programs for small firms: experimental evidence from Yemen, *Journal of Development Effectiveness, 9*(1), 1-14.

MLE (2013.) Types of evaluation designs. Measurement, Learning and Evaluation Project. (Retrieved from: https://www.urbanreproductivehealth.org/toolkits/measuring-success/types-evaluationdesigns)

National Council for Voluntary Organisations (NCVO), 2012, *Measuring Impact - A Guide to Resources*, available at http://www.ncvo-vol.org.uk/policy-researchanalysis/research/measuring-impact, accessed 16 January 2012.

Newman, J., Pradhan, M., Rawlings, L. B., Ridder, G., Coa, R., & Evia, J. L. (2002). An Impact evaluation of Education, Health and Water supply investments by the Bolivian Social Investment Fund, *World Bank Economic Review,* 16 (2), 241-274.

Owen, J. M. (1993) *Program Evaluation, forms and Approaches*, Allen and Unwin, St. Leonards, N.S.W., Australia.

Paxson, C., & Schady, N. R. (2002). The allocation and impact of social funds: Spending on school infrastructure in Peru, *World Bank Economic Review*, 16, 297–319.

Perkins, S. M., Tu, W., Underhill, M. G., Zhou, X., & Murrouy, M. (2000). The use of propensity scores on pharmacoepidemiologic research, *Pharmacoepidemiology and Drug Safety,* 9(2), 93-101.

Pitman, E. J. G. (1937). Significance tests which can be applied to samples from any population III. The analysis of Variance test, *Biometrika*, 29, 322-335.

Pitt, M. M., & Khandker, S. R. (1998). The impact of group-based credit programs on poor households in Bangladesh Does the gender of participants matter? *Journal of Political Economy*, 100(3).

Rao, V., & Woolcock. M., (2003). Integrating qualitative and quantitative approaches in program evaluation. *Qualitative and quantitative approaches*, Chapter 8:165-190.

Raudenbush, S. W., & Bryk, A. S. (2002). Hierarchical linear models: Applications and datas methods, *Thousand Oaks, CA: Sage publication.*

Ravallion, M. *Evaluating anti-poverty programs* (2005). Washington, D.C.: World Bank.

Ravallion, M., & Wodon, Q. (2000). Does child labour displace schooling? evidence on behavioural responses to an enrollment subsidy. *The Economic Journal, 110*(462). doi:10.1111/1468-0297.00527.

Ross, J. A., Hogaboam-Gray, A., Gray, P. (2004) Prior student achievement, collaborative school processes, and collective teacher efficacy. Leadership and Policy in Schools, 3(3), 163-188.

Rubin, D. B., & Thomas, N. (2000). Combining propensity score matching with additional adjustments for prognostic covariates, *Journal of American Statistical Association,* 95, 573- 585.

Scriven, M (1967). *The methodology of evaluation. In Curriculum evaluation.* Edited by Stake Chicago: Rand McNally.

Scriven, M. (1991) *Evaluation Thesaurus* n). (4th Edition) Newbury Park: Sage.

Stake, R. E. (1973) Programs evaluation: particularly responsive evaluation. In *Evaluation Models,* edited by G. f. Madans, M. S. Scriven and D. L. Stufflebeam Boston Khwer-Nijhoff.

Suvedi, M. and Kaplowitz, M. D. (2016.) Process skills and competency tools – what every extension worker should know – *Core Competency Handbook.* Urbana, IL: USAID-MEAS (available at: http://www.meas-extension.org/meas-offers/training).

Suvedi, M., and Vander Stoep, G. (2014.) Improving the monitoring and evaluation of agricultural extension services. Discussion paper. East Lansing, Michigan, U.S.A.: Department of Community Sustainability, Michigan State University.

Tashakkori, A., & Teddlie, C. (1998). Mixed Methodology, Combining Qualitative and Quantitative Approaches, *Sage, Thousand Oaks.*

USAID. (2011.) *Evaluation policy.* Washington, D.C., U.S.A.: Bureau for Policy and Planning.

Worthen, B. R., Sanders, J. R., & Fitzpatrick, J. L. (1997). *Educational evaluation: Alternative approaches and practical guidelines.* (2nd ed.). White Plains, NY: Longman.

Chapter 8

# The Future of Extension: A Global Vision

**Seema Kujur***
**Subhashree Sahu**
**Manjeet Singh Nain**
**and Sukanya Barua**
Division of Agricultural Extension, ICAR-IARI, New Delhi, India

## Abstract

Extension services are critical in bridging the gap between knowledge creators and knowledge users, facilitating the transmission of information, technology, and practices to enhance livelihoods and promote sustainable development. As we look to the future, we must examine the changing agricultural, technological, and socio-economic landscapes that will define the worldwide orientation of extension services. Future extension must address the recent technological developments, socio-economic factors, and environmental necessities. Technological innovation will significantly impact the future of extension services since digitization and data-driven decision-making will play a significant role. Information-based and communication-based technologies such as artificial intelligence, blockchain, and precision agriculture are emerging technologies that can transform the delivery of extension services. Furthermore, the process of urbanization, along with changes in population characteristics and shifting consumer preferences, will undoubtedly need a transition towards agricultural production systems that are more varied and offer more excellent value-added products, enriching the nutritional values of these products. Extension

* Corresponding Author's Email: lilyseema805@gmail.com

In: Global Horizons in Agricultural Extension Education
Editors: R. K. Doharey, Aman Verma, Satyapriya et al.
ISBN: 979-8-89530-446-4

programs must adjust adequately to meet the diverse requirements of smallholder farmers, women, and youth, as well as to retain them in agricultural activities for their livelihood security. Also, given the increasing severity of environmental problems, extension services must prioritize resilience, adaptation, and climate-smart behaviors to ensure a sustainable future. Extension services must play an important role in promoting conservation of agriculture and the sustainable use of natural resources to make the agricultural system resilient. To meet these challenges, the future of extension services necessitates a comprehensive and forward-looking strategy incorporating technological advancement, including nutri-smart agricultural practices, socio-economic inclusiveness, and environmental sustainability. Given the increasing severity of environmental issues, extension services must focus on building resilience, promoting adaptation, and encouraging climate-smart practices to secure a sustainable future. They play a crucial role in advancing conservation agriculture and ensuring the sustainable use of natural resources. To address these challenges, future extension services require a comprehensive, forward-thinking strategy integrating technological advancements, nutri-smart agriculture, socio-economic inclusiveness, and environmental sustainability. This approach is essential to fostering both agricultural resilience and long-term environmental health.

**Keywords:** extension service, farmer to farmer, nutri-sensitive agriculture, climate resilient, technology transfer, rural youth, rural women, information and communication technologies agri-preneurship

## 8.1. Introduction

Extension as a practice has a long history, spanning both ancient and contemporary times (Kassa & Temesgen, 2011). When "University Extension" was introduced in Britain in 1840, the term "Extension" was first used to refer to extending education beyond university walls. The Latin origins of the word "extension" refer to the idea of extending research findings, knowledge, and innovation to distant fields and places outside the purview of formal education, which is limited to institutions of higher learning. One could argue that Extension has existed since the dawn of human civilization and agriculture. Archaeological evidence suggests that individuals were exchanging information about improved agricultural practices and the Mesopotamian civilization during that time is a prime illustration of this; which was a triumph that changed agriculture and rural life in numerous world regions. Over time, extension has taken on various forms and functions (Temesgen and Kassa, 2011). With three components—adult education, technology transfer, and advising services—the concept of Extension was first proposed in the late nineteenth century in both the U.S. and the U.K. (Shinn et al., 2009). The two ancient universities, Oxford and Cambridge, started discussing how they could meet the rapidly industrializing urban area's rapidly expanding educational demands around 1850. The term "university extension" was not applied to anything other than a first practical attempt in 1867. Still, the activity swiftly gained traction, and by the end of the century, it was a well-established movement (Garforth et al., 1997). Ultimately, the extension has developed into a formalized public service that is funded heavily, employs many people, and is carried out by governments to enhance rural residents' capacity to accept new practices and technologies to adapt to shifting circumstances and societal demands (Kassa, & Temesgen, 2011). Extension education is a specific field of science that aims to educate, motivate, and alter people's behavior. It is pursued in many other areas, including agriculture. With ground-breaking initiatives in nations like the United States, Germany, and Japan, extension has its roots in the late 19th and early 20th centuries. Global extension systems saw substantial change in the 20th century due to the necessity of bringing agricultural technologies and information to rural areas. Extension services have evolved into formalized public systems, heavily funded and employing large workforces, with governments implementing them to help rural communities adopt new practices and technologies. This adaptation enables rural residents to meet changing societal demands and conditions (Kassa & Temesgen, 2011). Extension education is a scientific field aimed at educating, motivating, and influencing behavioral changes in people. While it applies to many areas, its impact has been particularly significant in agriculture. The concept of extension originated in the late $19^{th}$ and early $20^{th}$ centuries, with pioneering initiatives in countries like the United States, Germany, and Japan. During the 20th century, global extension systems underwent significant transformations to meet the growing need for disseminating agricultural information and technologies to rural populations.

The term "extension" starts from university extensions, which were in use in nineteenth-century England. It was first applied to challenges related to agriculture and a few other areas of societal development. Subsequently, it began serving as a tool in different fields of development. The idea and tenets of extension are increasingly being applied to various development domains, both urban and rural. For example, the development of micro and small enterprises, marketing, health, and other sectors successfully utilizes extension principles and practices in Ethiopia (Kassa & Temesgen, 2011). In the 21st century, society shifted from emphasizing local governance within a sense of place to a mass society. Extension is reacting increasingly to the learning requirements of both communities of interest and location as we go into the information age, which tends to focus on communities of interest. Everyone needs to contribute to stay relevant in this changing environment. Society changed from emphasizing local government within a sense of place to becoming a mass society. As we go into the information age, which tends to focus on communities of interest, extension responds more and more to the learning requirements of both locations and communities of interest. If individuals wish to remain relevant in this ever-changing environment, they must all participate (Bull et al., 2004). The relevance of extension in the future depends on whether cooperative extension adheres to the values that have guided it over the past 100 years and whether it can and will remain incredibly relevant. Suppose Cooperative Extension cannot produce a significant economic benefit for the public good considering changing bureaucratic goals. In that case, its supporters will grow less, and the program will lose relevance (Smith & Smith, 2007).

## 8.2. The Birth of Modern Agricultural Extension Services in the World

The first modern agricultural extension agency was established in response to a potato blight outbreak that struck Europe in 1845. The effects were nasty in Ireland, where most people lived in rural areas and relied on potatoes for food; "the potato famine" persisted until 1851. Meanwhile, the French government entirely funded establishment of the country's first agricultural extension organization in 1879. The establishment of similar initiatives in the United Kingdom was significantly influenced by the expansion of agricultural education and extension efforts throughout continental Europe. In Japan, efforts to extend agriculture had also begun before 1900. New administrative frameworks and modernizing policies were implemented during the Meiji Restoration of 1868. In the mid-1870s, two agricultural colleges were established with teachers from the West, primarily from Europe. A national department of agriculture was founded in India in 1866 in response to the famine that struck Orissa. The growth of agricultural education and extension work in continental Europe was to have a strong impact on the emergence of comparable activity in the United Kingdom. Agricultural extension work had also started before 1900 in Japan. After the Meiji Restoration in 1868, new modernizing policies and administrative frameworks were implemented. Two agricultural colleges were founded in the middle of the 1870s, employing Western (mostly European) professors. Following the famine in Orissa in 1866, India established a national department of agriculture. The Indian government quickly decided to establish ministries in each province. However, it wasn't until 1905 that a central government mandate forced every province to appoint a full-time director of agriculture, whose job it was to establish research and demonstration farms for agricultural reasons and staff them with guidance from farmers (Mook, 1982).

## 8.3. Paradigm Shifts of Agricultural Extension

In the past, top-down technology transfer was the dominant extension technique. That paradigm failed to provide meaningful benefits to many users, while the resource-rich clients received the benefits. As a result, new thoughts and approaches were necessary.

### 8.3.1. Technology Transfer (Persuasive + Paternalistic)

The Transfer of Technology approach, which gained traction in the 1960s, was a widely accepted model of extension services. It is most effective when dealing with straightforward solutions that can be easily communicated to a large audience, often called a Science-push strategy ("The modern valuation of agricultural extension," 2019). This method, reminiscent of colonial times, resurged in the 1970s and 1980s with the establishment of systems like "Training and Visit" across Asia. Technology transfer follows a hierarchical structure in this paradigm, providing farmers with specific recommendations for adopting practices. Initially championed by Land Grant Universities in the U.S.A., the University of Illinois now stands out for implementing this approach (Babadoost, 2018). Notable examples include the Training and Visit system (T&V), Farming Systems Research (F.S.R.), and Farmer Field Schools (FFS) (Wijeratne & De Silva, 2024; Bachano & Jimma, 2019).

### 8.3.2. Advisory Work (Persuasive + Participatory)

This paradigm is still in place today, as private consulting firms or government agencies reply to questions from farmers with technical recommendations. It can also take the shape of initiatives run by non-profits and donor organizations that promote pre-packaged technology through participatory methods (Bachano & Jimma, 2019). Demand-driven extension is a common term to describe this new paradigm in extension. According to Akinnagbe and Ajayi (2010), demand-driven extensions are frequently associated with the idea of privatization and a shift away from free public services.

### 8.3.3. Human Resource Development (Educational + Paternalistic)

The early days of extension in Europe and North America, when colleges trained rural residents who were too impoverished to enroll in full-time classes, were dominated by this paradigm. It is still practiced today in college outreach programs all around the world. Although top-down teaching techniques are used, students are expected to choose how to apply the knowledge they have learned. (Jimma & Bachano, 2019). Regarding the human resource paradigm, a paternalistic and pedagogical strategy is employed. This paradigm resulted from colleges performing extension work for people who cannot afford to attend, hence supplying education and knowledge. Additionally, it gives the pupils the freedom to decide for themselves using the facts they have learned. However, this paradigm still uses a top-down approach; thus, it is still paternalistic (Wijeratne & De Silva, 2024).

### 8.3.4. Facilitation for Empowerment (Educational + Participatory)

This paradigm uses farmer-to-farmer exchanges and experiential learning techniques. Through participatory processes, knowledge is acquired, and participants are urged to exercise independent judgment. Farmer Field Schools (FFS) and participatory technology development (P.T.D.) initiatives are the most well-known examples in Asia. The final paradigm, facilitation for empowerment, is participatory and instructional in character. It is participatory because it incorporates farmer-to-farmer communication and an interactive process to educate farmers, and it is educational because it offers new knowledge, practice, and skills on how to boost output and safeguard the environment (Wijeratne & De Silva, 2024).

## 8.4. The Extension System in India

The Indian extension system has a long history from the first half of the 1900s. The nation's extension services have been essential in helping farmers adopt modern farming methods, provide technical support, and spread agricultural knowledge. Agricultural extension was implemented by the British colonial authority in the 20th century to tackle the issues that Indian agriculture faced, including low productivity, pests, and illnesses. Formal extension services began with establishing agricultural departments, demonstration farms, and agricultural schools (Behera et al., 2024). Among the many forms of government investment in agriculture, agricultural research, education, and extension are considered the most important for boosting farm production and increasing revenue (Nedumaran & Ravi, 2019). "As a social institution, extension needs to change with society all the time." Though it is impossible to predict the future of global extension, one could argue that its philosophy has always been to change to meet the requirements of its intended audience, whether or not they are aware of them. Therefore, rather than being reactive, extension needs to be proactive and flexible in solving issues facing agriculture to ensure its continued relevance. At the local and national levels, networking and forging solid relationships are two things that extension has always been adept at (Davis et al., 2018).

### 8.4.1. Agricultural Extension Approaches

There are 8 significant approaches commonly accepted worldwide based on the various problems to which the approach must be applied (Mallick et al., 2023). There are 8 major approaches commonly accepted worldwide based on the various problems to which the approach must be applied (Mallick, et al., 2023).

#### *8.4.1.1. General Agricultural Extension Approach*

This strategy includes extension plans, projects, and ministry-based initiatives, similar to public expansion. It claims that if knowledge and technology are appropriately distributed throughout the farming community, farming might become more productive and productive. The success of this strategy is gauged by the rate at which suggested technologies are adopted, as well as the growth in productivity and output at the national level. The best illustration of a generic extension strategy is the extension system of the State Agriculture Department of India. This strategy can accomplish several important goals, including implementing the national program, a thorough extension network down to the village or block level, and effective communication between different ministries. This strategy has several drawbacks, including its top-down methodology, lack of two-way communication, inadequate infrastructure for meeting farmer demands, incorrect information distribution, limited adoption of technology, and high expense. A prime example of a generic extension strategy is seen in the State Agriculture Department of India's extension system. This approach helps achieve key objectives, such as rolling out national programs, establishing a widespread network reaching villages and blocks, and fostering coordination between ministries. However, it also has notable shortcomings, including a top-down approach, insufficient two-way communication, limited infrastructure to address farmers' needs, dissemination of inaccurate information, slow adoption of technology, and high costs.

#### *8.4.1.2. Commodity-Based Extension Approach*

Cash crops grown for commercial export, such as tea, coffee, sugar, spices, etc., are covered by this approach. The commodity-based extension advice system involves statutory commodity boards, such as the boards for tea, coffee, spices, tobacco, rubber, profit-oriented businesses, farmer-producer organizations, or individual entrepreneurs. The extension system handles many tasks, such as marketing the product and providing feedback for adopting new technologies. Thus, the plan offers extension services, including inputs, credits, consulting services, training, quality management, and marketing strategy while focusing on a single crop under one administration. The targeted crop's entire yield is considered when determining the success of this extension strategy. The approach's main benefits include a high rate of adoption, technology that is appropriate for the agroecological zone, timely messages that are tailored to the needs of the farmers, a strong connection between input suppliers and marketing agents, a low farmer-to-agent ratio, and easy access to additional extension services like credit training and inputs. One of the disadvantages is that decision-making about the commodity can lead to system failure.

#### *8.4.1.3. Training and Visit (T&V) System of Extension*

This approach encourages farmers to grow more of a specific crop and creates communication channels between extension agents, researchers, and staff. The T&V approach has many advantages, including more field extension employees, a more professional approach to extended services, up-to-date extension professionals with access to the most recent agricultural knowledge and technology, and logistical support. Notable drawbacks include the top-down methodology, which precludes a two-way exchange of information. Bureaucratic decisions suppress the actual needs of farmers. T&V exclusively focuses on transferring technology—rather than generating it—because it is a supply-driven strategy. The system is less flexible and more complex due to the lack of low-cost technology.

#### *8.4.1.4. Participatory Extension Approach*

This strategy is founded on self-mobilization, where farmers assess and evaluate their significant issues and look for workable solutions based on their requirements. This method's primary feature is that the farmers cultivate their land sensibly. In contrast, this method requires fewer resources. Any extension project must be implemented with group meetings, demonstrations, and individual and group travel as essential components. The extension approach has a high degree of usefulness and efficacy. Rather than having authorities thrust any agenda upon them, farmers identify their concerns. The main drawbacks are that the center has no control over the program's administration or access and no long-term commitment to keep the program going. Important Participatory approaches are Participatory Rural Appraisal (P.R.A.), Rapid Rural Appraisal (R.R.A.) and Participatory Action & Learning Method (PALM) etc.

#### *8.4.1.5. Project Approach*

This strategy concentrated on a time and place specific program to raise rural residents' standards of life and enhance their production situation. A project management team, project allowances for field workers, improved transit options, equipment availability, lodging, and the provision of foreign consultants for local staff are all included in the implementation plan. This approach's success is measured in the short term. Project-based extension initiatives are shown by agricultural development programs such as the Intensive Agricultural District Programme (1960), Operational Research Project (1974), Lab to Land Programme (1979), IARI-Post Office linkage extension model (2009), and so on.

#### *8.4.1.6. Farming System Approach/Farming System Research and Extension (FSR&E)*

Extension initiatives were frequently turned down because the suggested technology did not meet the needs of the local population. Farming System Research and Extension, or FSR&E, was created as a new extension strategy to address the shortcomings of the top-down extension paradigm used in the 1970s. Its goal is to spread technology that helps the local farmers meet their needs. Using the FSR&E approach, specialized field staff members establish a link between farmers, research, and extension to disseminate technologies spanning various scientific fields. A group of multidisciplinary specialists classifies issues pertaining to farmers' requirements and offers training, field experiments, and demonstrations in farmers' fields. The degree to which the technologies are adopted and maintained throughout time is a standard indicator of the approach's effectiveness. The FSR&E approach's primary benefits are its need-based, extension-research-linked extension initiatives. The technique is non-functional due to limitations such as insufficient input availability, ineffective marketing strategy, inadequate cooperation and participation from relevant departments, and indifferent officials. The "Bio-resource complex" of the University of Agricultural Science, Bengaluru, the "Rural demonstration and training center" of the village Padhri Kalan, Amritsar, Punjab, the "Roopa farm" of Mohinder Singh Grewal of Punjab, and other examples are examples of successful F.S.R.E. models.

#### *8.4.1.7. Cost-Sharing Approach*

This strategy seeks to boost farm output and help farmers learn new skills for personal development. People's ability and desire to share a portion of the project's costs determines success. Some noteworthy benefits include developing community-based programs and techniques based on needs, involving the local population to raise adoption rates, boosting local confidence, and reducing the financial strain on the government by having them share some of the costs of the services. This strategy overcomes obstacles such as marginal and small farmers' reluctance to cover expenses, the government's occasional refusal to pay, and communities' objections to management choices since they act like owners. I.T.C., Agri-clinic, and M.S.S.R.F. are a few instances of cost-sharing approaches.

### *8.4.1.8. Educational Institution Approach*

The strategy primarily concentrates on extension operations carried out in agriculture colleges and universities. The technical expertise that agricultural university faculty members possess substantially influences farmers' understanding of scientific agriculture. The degree to which farmers attend and participate in agricultural extension events is a gauge of this strategy's effectiveness. Authentic problems from the field are provided to the extension fraternity, allowing scientists and field extension personnel to continue their working relationship, ensuring effective use of human resources, and lowering the financial burden of providing extension services on the national government are just a few of the approach's unique advantages. This technique has the following limitations: the program is solely academic. As a result, it causes gaps between subject matter experts and field workers; from the farmers' viewpoint, the lectures are less valuable and practical. Extension education institutions (E.E.I.s), the IARI-Post office linkage extension model, the IARI-VO partnership model of the Indian Agricultural Research Institute, and the Punjab Farmers Club Extension strategy are some of the most compelling examples of this extension approach.

## 8.5. The Future of Extension

The future of extension services globally is characterized by innovation, adaptation, and collaboration to address the evolving needs and challenges of agriculture and rural development in a rapidly changing world with the following ongoing broad areas of extension:

1. Social Learning-Based Extension/Farmer-to-Farmer Extension
2. Nutrition-Led Extension/Nutri-Sensitive Agriculture
3. Information and Communication Technology-based Extension
4. Agri-preneurship Extension
5. Climate Resilient Extension
6. Gender/Women Inclusive Extension Services
7. Extension efforts in Promoting Youth Engagement and Employment in Agriculture

### 8.5.1. Social Learning-Based Extension/Farmer-to-Farmer Extension

Scarborough et al., (1997) define a farmer-to-farmer extension, or F2FE, as "the provision of training by farmers to farmers, often through the creation of a structure of farmer promoters and farmer trainers." Different titles for this strategy, such as "farmer-trainer," "lead farmer," "farmer-promoter," and "community knowledge worker," may allude to distinct functions (Franzel *et al.*, 2015). It began in Guatemala in the 1970s and eventually spread to Nicaragua, Mexico, and Honduras in the 1980s. The farmer-to-farmer extension model has emerged due to agricultural extension's shift towards community-centric, farmer-centered, and participatory techniques that strengthen rural development through a social learning lens.

When stakeholders, particularly the community, are actively involved in the model's design, implementation, and evaluation, the farmer-to-farmer method may be more successful (Ochieng et al., 2022). These days, it is widely used in many Latin American, Asian, and African nations, undergoing several modifications and adjustments (Weinand, 2002; Meena et al., 2016). Adoption of technology is crucial for smallholder farmers in the agricultural sector. Farm-to-farmer training increased production and the uptake of new technology.

According to Nakano et al. (2018), untrained farmers have surpassed trained farmers in yield and profit. The strategy suits many target groups, including young people, women, and the underprivileged. It is beneficial for expanding women's access to extension services and the percentage of female extension providers. More female farmer-trainers than female front-line extension employees can be found in many locations. For instance, women comprised fewer than 5% of extension employees but around one-third of volunteer farmer-trainers in Uganda's East African Dairy Development Programme (Meena et al., 2016). The principles of F2F Extension Model are given below. This model can assist in the development of effective, farmer-centered extension systems, as well as the empowerment of farmers as change agents in their communities to improve livelihoods.

1. *Selection of Lead Farmer*: The process of selecting lead farmers, monitoring, and assessment involves several stakeholders, including farmers, local institutions, village leaders, key informants, etc. This aids in increasing the programs' accountability to the farming community they serve.

2. *Communication*: Lead farmers are members of the community and are regarded as basic members of it. They speak in plain, regional languages with their fellow farmers. They have a greater awareness of regional customs, agricultural methods, and farmers' needs, interests, and resources.
3. *Skills and Interest:* The selection of lead farmers is contingent upon their aptitude and propensity for information exchange. Instead of focusing on a single criterion like large farms or farming experience, the goal should be to increase intrinsic learning competencies.
4. *Linkages:* Lead farmers require strong ties to and assistance from research institutions, development organizations, and extension organizations that provide them with training and support. Lead farmers are not a replacement for the pluralistic extension systems now in place but rather their complement.
5. *Facilitation:* Promoting organizations and regional establishments must take the initiative to guarantee that women and men alike take on the role of lead farmers.
6. *Reference materials*: Lead farmers should have access to adequate and basic reference materials.

### 8.5.2. Nutrition Led Extension/Nutri-Sensitive Agriculture Extension

Malnutrition remains a significant development concern in most South Asian countries, even after several decades of the green revolution. Most of the region's malnourished population resides in India (Kumari & Junuthula, 2023). The current food crisis is a symptom of a more extensive breakdown in the global food system. The triple burden of malnutrition—undernutrition, overnutrition, and deficiencies in micronutrients—is something that the globe is currently grappling with. A significant public health concern is the triple burden of malnutrition. According to the World Bank Group Joint Child Malnutrition Estimates (2023), N.I.C.E.F., WHO, and World Bank combined report, stunting affected an estimated 22.3% of children or 148.1 million, overweight affected 5.6% of children or 37 million, and wasting threatened the lives of an estimated 6.8% of children under the age of five worldwide in 2022. Therefore, it is crucial to comprehend the connections between nutrition and agriculture.

The agricultural extension paradigm has shifted from being driven by production to being led by markets, and it is currently in the transitional phase of adopting the new concept of nutrition-led extension (farm to plate). We must implement educational programs to guarantee people's nutritional security by educating them about the worth of nutrients and showcasing value-added items that they can incorporate into their diet." The goal of nutrition-sensitive agriculture is to produce safe, healthy, culturally relevant, and reasonably priced food in sufficient quantities and quality to sustainably meet the population's nutritional needs (Uccello et al., 2017). By producing a variety of safe, nutrient-rich foods, generating income to help with access to healthcare, lowering water source contamination, and utilizing labor-saving technologies, among other things, nutrition-sensitive agriculture and food systems help to improve health outcomes (Uccello et al., 2017). Several efforts have been made to connect agriculture and nutrition through extension:

### *8.5.2.1. Community Agri-Nutri Security Centres*

The Division of Agricultural Extension, Indian Agricultural Research Institute (I.C.A.R.) in New Delhi, India, undertook a DST project titled "A Nutrition-led Extension Model of Community Agri-Nutri Security Centres (C.A.N.S.C.s) for Nutrition Security." Several awareness campaigns were carried out as part of this initiative to combat malnutrition and promote Good Nutrition Practices (G.N.P.) (Sangeetha et al., 2016). For this very purpose, ICAR-IARI has developed the Agri-Nutri Smart Village Model or A2N model as illustrated in Figure 8.1.

### *8.5.2.2. Agri-Nutri Smart Villages*

To transform Indian villages into nutritionally self-sufficient units by utilizing local resources and providing appropriate nutrition education, a link between agriculture and nutrition must be established. As a result, novel ways and models are required to provide food from farm to plate. The suggested model, "Agri-Nutri smart villages" has the following six components.

A. *Nutri-Farming System*

1. *Cultivation of Traditional Crops:* Traditional nutritious crops such as coarse grains, jowar, bajra, and pearl millets have lost their position in kitchens despite their high nutritional and mineral content. Traditional crops must be incorporated into village farming patterns, now called Nutri-cereals.

2. *Cultivation of Nutri-rich Varieties:* Most farmers cultivate the same crop year after year. The hour needs to vary cropping patterns and include nutrient-rich crop varieties. ICAR-IARI, New Delhi, often known as the Pusa Institute, has a basket full of nutrient-dense species.
3. *Fruit and Vegetable Cultivation:* Farmers who focus solely on rice-wheat farming patterns must be encouraged to grow fruits and vegetables for at least domestic consumption.
4. *Forage and Fodder Crop Cultivation for Nutritional Security*: Farmers should be properly trained to arrange their animal feed, such as berseem, lettuce, and other crops. Farmers must be advised to keep milk for family members rather than marketing it whole.

B. *Nutri-club:* At the village level, adults, women, youth, and children can form a Nutri-club. This group mainly organizes weekly rallies, exhibitions, nutritional campaigns in schools and Anganwadis, and Nutri-quizzes for kids, teens, and mothers. Similar to Media Forums, which originated in Canada and give farmers access to mass media and one-on-one communication regarding agricultural techniques, separate Nutri forums can be set up at the village level.

C. *Agri-Nutri Education*: E-modules can disseminate excellent dietary habits and nutritional information throughout villages. Agricultural research institutes, Krishi Vigyan Kendras (K.V.K.), and state agricultural universities, in collaboration with nutrition specialists, will play an essential role in generating content for e-learning modules and messaging addressing agriculture and nutrition.

D. *Agri-Nutri Entrepreneurship:* Entrepreneurship development is another area where nutrition should be improved. The youth of villages must be trained to prepare nutritious cuisine. ICAR-IARI has developed a number of nutrient-dense post-harvest items that may be produced and delivered at the village level for a cheap cost.

E. *Agri-Nutri Capacity Building*: Rural youth capacity building can be achieved through behavioral change interventions such as trainings and seminars. Through a number of initiatives, state agricultural colleges, state governments, and other agricultural institutions in India can carry out this skill gap analysis in the area of Agri-nutrition. Research institutes, K.V.K., and A.T.M.A. might collaborate to develop and provide training.

F. *Nutri-kitchen Garden*: The bulk of farming families have largely fallow backyards. It is possible to turn the backyard into a kitchen garden where fruits, vegetables, and flowers can be produced for household use. Farmers can also cultivate mushrooms and maintain apiculture. The village's nutria-club and a resource person from the research institute can help organise a demonstration of the kitchen garden's technical know-how.

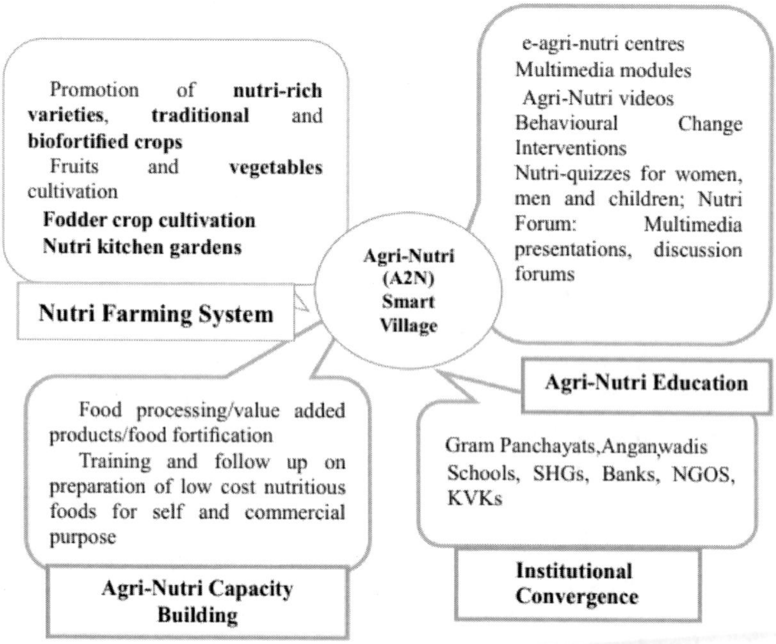


**Figure 8.1.** Agri-Nutri Smart Village Model (Sangeetha et al., 2018).

### 8.5.3. I.C.T. Based Extension

Agricultural extension has been at the forefront of providing farmers with sufficient information to boost productivity over the years. The degree of agricultural output depends mostly on how well farmers are informed about these techniques and information (Ezeh, 2013). Incorporating digital technologies into conventional processes has caused a notable revolution in the agriculture industry. The advent and extensive integration of digital technology have fundamentally transformed the agricultural extension services sector, promoting enhanced efficacy, reach, and durability. A new era of agriculture extension services has been brought about by the widespread use of digital technologies, including cell phones, the internet, data analytics, and sensor-based devices (Singh et al., 2018). I.C.T. is a crucial tool that supports extension efforts in the current context and will provide rural farmers and farm women with new information services that will enable faster and more extensive information sharing. For the farming system to function sustainably, access to these I.C.T. tools for information are essential (Jha, 2021). Knowledge is essential for smallholders, resource-poor farmers and producers struggling to make ends meet while generating money from their meager resources.

Information and communication technology (I.C.T.) in agriculture makes it easier for researchers, exporters, extension services, and farmers to share knowledge within and beyond different agriculture networks (Paroda, 2010; Kafura et al., 2016). Through radio shows, films, and other ICT-based techniques, public and private actors working in rural areas of developing nations have expanded their repertoire of knowledge-dissemination strategies during the past 20 years. Due to a shortage of funding and personnel to physically visit with all farmers and communities, public extension systems struggle to contact all farmers. Another significant issue is that, as farming becomes more and more focused on the market, farmers are asking for more specific and varied information. Even in agriculturally homogeneous areas, farming systems vary concerning the kind of crops grown, inputs used, labor-to-machinery ratio, and quality standards applied. Information and expertise must be customized to each farmer's needs to be genuinely effective (Barber et al., 2018).

#### *8.5.3.1. Expert System (E.S.)*

An expert system is a software program that mimics the actions of one or more human experts. Expert systems are a classic use of artificial intelligence primarily focused on a particular issue domain. "Using pre-established conditions found in the software program, the expert system is utilized to act as though it were a human expert to answer the problem. Expert systems typically have three main components: a knowledge base, an inference engine, and a user interface." IF-THEN rules are a way to represent the experience and expertise of a human expert (Rani et al., 2011).

### *8.5.3.2. Crop Management Expert Systems*

This category of E.S. includes advising systems focused on crop management. These systems often aim to provide a comprehensive and integrated decision-support strategy covering most crop growth.

- *E.S.I.M.:* An expert system for irrigation management was created to make choices regarding water management in an irrigation project. The Mae-Taeng Irrigation Project in northern Thailand had an irrigation management issue. Thus, the technology was used to solve it. Srinivasan *et al.* (1991) developed an interactive system designed to be user-friendly.
- *C.I.T.E.X.* Expert System: An expert system for producing oranges is called C.I.T.E.X. It offers services for farm evaluation, scheduling of fertilization and irrigation, diagnosis and treatment of disorders. The Central Laboratory of Agricultural Expert Systems (CLAES), located in Egypt, developed the system (Mishra *et al.*, 2014). The expert system fields yielded an approximate production increase of 28.7% on average compared to the control fields' yield. According to Mahmoud *et al.* (2008), the average cost of cultivating the E.S. fields is 6.5% lower than that of the control fields.
- *GRAPE:* It is an E.S. for viticulture in Pennsylvania. Pennsylvania State University developed this E.S.E.S. in collaboration with Texas A&M University to meet the advisory needs of grape farmers. This approach makes suggestions for grape farmers on pest management (insect, disease, and weed control), fertilization, pruning, and site selection. On the Macintosh microcomputer, the programming environment contained a rule-based shell known as Rule Master.
- *C.O.M.A.X.:* It is an expert in cotton crop management. The system employs a combination of expert-derived rules and results generated by the cotton crop simulation model G.O.S.S.Y.M. C.O.M.A.X. calculates the most effective irrigation, fertilization, defoliant, and cotton boll opening method. Sensors in cotton fields automatically send weather conditions to the system, and C.O.M.A.X. updates its recommendations daily. C.O.M.A.X. was tested on a big farm and showed outstanding results in lowering unit production costs (Lemmon, 1986; Mishra et al., 2014).
- *Tomato Expert System*: Babu et al., (2010) created a web-based tomato crop expert information system in India. This expert system is

divided into the tomato information system and the tomato crop expert system. The information system provides the user with static information about various species, diseases (including viruses), pests of tomato fruits and plants, their symptoms, preventions, and chemical controls. This tomato crop information expert system deals with diverse tomato crop varieties and identifies numerous diseases and pests that commonly affect tomato crops based on symptoms (Rani et al., 2011).

- *Rapeseed-Mustard Expert System*: Vinod et al., (2008) created an image-based rapeseed-mustard disease expert system in India. This expert system efficiently diagnosed and controlled economically essential diseases such as Alternaria blight, white rust and white rot, downy mildew complex, powdery mildew, and rapeseed-mustard white rot. Extension professionals, researchers, and farmers can utilize the system to diagnose and manage rapeseed-mustard infections. The user may simply identify the ailment using photographs of symptoms and a text explanation (Kumar et al., 2016; Rani et al., 2011; Kumar et al., 2008).
- *Rice Expert System*: In India, Sarma et al., (2010) created an expert system to diagnose and manage diseases in rice crops. The designed technique is meant to diagnose common illnesses that affect rice plants. A rice expert system is a computer program typically consisting of a knowledge base, inference engine, and user interface. This expert system supports many components, including a decision support module with interactive user interfaces for diagnosing specific disease signs.

### *8.5.3.3. Crop Disease/Insect-Pest Diagnostic Expert Systems*

1. *V.E.G.E.S.:* It is a multilingual expert system that diagnoses pests, diseases, and nutritional disorders in six greenhouse vegetables: pepper, lettuce, cucumber, bean, tomato, and aborigine. It provides a diagnosis based on a brief description of the damaged plant's exterior appearance. It then proposes a solution to the problem (such as fertilizer, modification, fungicide, or pesticide sprays).
2. *POMME:* An expert method for pest management in apple orchards. It was created in Virginia to aid in managing illnesses and insect pests on apples. This system informs growers on fungicides, insecticides,

freeze, frost, drought damage, non-chemical care choices, and disease modeling data. POMME was one of the first expert systems to incorporate the expert's decision-making process while advising producers on disease management decisions.

3. *D.D.I.S.:* It is a distance diagnostic and identification system created by the University of Florida. The technology allows users to upload digitized samples collected in the field for rapid diagnosis and identification of plants, illnesses, insects, and animals. D.D.I.S. allows agricultural extension professionals and specialists to communicate knowledge about plants, insects, and diseases. The method generates a digital image library containing site, crop, and pest or disorder information that can be utilized in teaching programs, assisted diagnosis, and data mining.
4. *C.A.L.E.X.:* It is an expert system for diagnosing peach and nectarine problems. The system analyses 120 peach abnormalities in California, including insects, illnesses, and cultural issues. The user begins a session by identifying the location of the plant where the problem occurs. The Expert System relies on the "Certainty Factor" to reach decisions.
5. *COUNSELLOR:* It was created in England by I.C.I. Agrochemicals to control insects and illnesses in wheat. It was created on a D.O.S. machine using the expert system shell Savoir. This shell is unique in that it leverages evidence networks to forecast disease risk and optimize treatment suggestions. COUNSELLOR provides a cost-benefit analysis for the therapy recommendation.

#### *8.5.3.4. Internet of Things*

The Agricultural Internet of Things (IoT) has introduced new innovations to agricultural productivity. It increases agricultural productivity, improves product quality, lowers labor costs, raises farmer income, and achieves true agricultural modernization and intelligence (Xu et al., 2022). Kevin Ashton created the phrase Internet of Things (IoT) in 1999. The Internet of Things (IoT) refers to the ever-expanding network of physical items with an Internet Protocol (I.P.) address for Internet access and the communication between these objects and other internet-connected devices and systems.

#### *8.5.3.5. Advantages of Application of IoT Technology in Agriculture*

Data can be used by farmers to create plans, manage and carry them out on their own, make necessary course corrections, and revise plans. Farmers can plan, schedule, and plan their crops without relying on the state. Better judgments may be made by farmers when it comes to when to plant, water, feed, protect, or harvest their crops. IoT technology assists in offering farmers solutions in several ways, such as lowering damage gaps while preserving high-yield output, monitoring environmental safety, preserving climatic conditions, and boosting the farming industry's profitability. Additionally, the farming process and produce have lower carbon footprints due to IoT technology (Mohapatra et al., 2021; Tzounis et al., 2017).

### 8.5.4. Agri-Preneurship Extension

Entrepreneurship development in agriculture has emerged as a significant field of study, policy, and development initiative, with the viability of small and marginal-scale farming dampening owing to rising cultivation costs, declining market returns, and degradation of natural resources (Ray et al., 2022). As one of the main ways to generate revenue, entrepreneurship is crucial to the economic success of every nation. Most people in India still rely on agriculture as their primary source of income; thus, encouraging farmers to switch from subsistence to commercial farming and fostering an entrepreneurial spirit in the sector is essential. Numerous economic issues that affect India in general and agriculturists in particular might be resolved through agri-preneurship. It can make a big difference in rural development. Global extension services can improve agri-preneurship (Murali & Thomas, 2022). Extension strategies that are used to promote agriculture-related entrepreneurship among youth and women in tribal and rural areas include gender mainstreaming, bottom-up planning, identifying the area for agri-preneurship development, training, and skill development, capacity building, promotion of community-based organizations, convergence and public-private partnerships, and gender mainstreaming (Lakshmi, & Naidu, 2022).

#### *8.5.4.1. The Role of Agri-Preneurship in the Development of the Indian Economy*

Agri-preneurship is a transformative strategy that can improve rural livelihoods, spur sustainable economic growth, and advance national well-being. The following are some significant ways that agri-preneurship advances the development of the Indian economy:

1. *Enhanced Agricultural Productivity*: Agri-preneurs use cutting-edge procedures, effective resource management strategies, and contemporary technologies. Reduction of post-harvest losses, improved quality of produce, and higher crop yields are the results of this, which raises agricultural production.
2. *Diversification and Value Addition*: Agri-preneurship encourages farmers to diversify their goods and pursue value-added businesses like food processing, packaging, and branding.
3. *Job Creation and Rural Employment:* Agribusiness creates job prospects in rural areas. When farmers become agri-preneurs, they generate employment in other sectors like marketing, transportation, processing, and working on their own farms.
4. *Food Security:* Agribusiness can improve food security by advancing agricultural techniques and guaranteeing a steady supply of high-quality produce. To feed India's expanding population, this is imperative.
5. *Environmental Sustainability*: Agri-preneurship encourages environmentally friendly farming methods that preserve biodiversity, cut down on chemical use, and preserve natural resources. This is consistent with India's dedication to protecting the environment and lessening the effects of climate change.
6. *Market Linkages and Trade:* Because they focus more on the market, agribusiness owners can better connect farmers and consumers. Agribusiness endeavors enable farmers to connect directly to markets, streamline supply chains, and engage in domestic and foreign trade.
7. *Innovation and Research*: Agri-preneurship promotes agricultural innovation and research. Agribusiness entrepreneurs frequently experiment with new crops, methods, and technologies, resulting in locally relevant innovations that can be expanded for more comprehensive agricultural development.

India's agricultural environment might be drastically changed by agripreneurship, which would also support long-term, sustainable economic growth. It's critical to consider variables like market demand, resource availability, technical improvements, and legislative assistance when determining regions where agripreneurship can flourish. Arumugam and Manida (2023) identified several key sectors in agriculture: livestock farming, high-value crops, and horticulture, waste management and bioenergy, contract farming, agri-export, greenhouse farming, organic farming, precision farming and the Internet of things, aquaculture and fisheries, vertical farming and hydroponics, agri-tourism, agri-education and training.

#### *8.5.4.2. Government Interventions in Strengthening Agri-Preneurship in India (Gangwar & Jadoun, 2023)*

1. *Make in India:* The Government of India launched the "Make in India" initiative in 2014 intending to promote domestic development, manufacturing, and assembly. The goal of the policy approach was to create an environment that was advantageous to investments, skill development, and job creation.
2. *Dairy Entrepreneurship Development Scheme (D.E.D.S.):* The Department of Animal Husbandry, Dairying and Fisheries (Ministry of Agriculture) launched the "Dairy Entrepreneurship and Development Scheme" in 2006. The scheme's purpose was to encourage entrepreneurship in the dairy industry and improve the lot of dairy farmers.
3. *Agri-clinics and Agribusiness Centre*: The "National Institute of Agricultural Extension Management" (MANAGE) in Hyderabad, which assists in implementing these projects, and the Government of India established the agri-clinics and agribusiness centers. The plan aims to enhance the current extension network to support input supply and services and accelerate the transfer of agricultural technologies.
4. *Rastriya Krishi Vikas Yojana (R.K.V.Y.):* To increase farmer income and provide job prospects for the rural people, the Indian government introduced the "Rastriya Krishi Vikas Yojana" in 2020 (Yadav et al., 2022). The program focuses on helping farmers become more innovative and entrepreneurial.
5. *Paramparagat Krishi Vikas Yojana:* The Indian government launched the "Paramparagat Krishi Vikas Yojana" in 2015 to promote organic farming nationwide. Thus, farmers were invited to join clusters or groups to practice organic agriculture throughout a vast region of agricultural land (Reddy, 2018).
6. *National Beekeeping and Honeybee Mission (N.B.H.M.):* To accomplish the "Sweet Revolution," the Ministry of Agriculture and Farmer's Welfare launched the program in 2019–20. Farmers and beekeepers received scientific training to advance scientific beekeeping and encourage agripreneurship in the industry.

### 8.5.5. Climate Resilient Extension

Climate change and its associated disasters and extreme events are significant factors contributing to food insecurity across all dimensions. Globally, these variables undermine existing efforts to end hunger and undernutrition by raising the likelihood of food insecurity to previously unheard-of levels. Given how extremely vulnerable agriculture is to climate change, challenges will worsen. The production of crops, cattle, fishery, and forestry is negatively impacted throughout the region by reduced rainfall, more frequent extreme weather events, and hotter and shorter growing seasons. This has major implications for the security of food and nutrition. Building climate resilience in a way that ensures a growing population can be fed sustainably without further depleting natural resources is necessary to prevent such trends.

This entails assisting people in dealing with current changes, adapting their livelihoods, and improving governance mechanisms to manage future risks better. Long-term plans and interventions based on agro-ecological knowledge are also required to enable smallholder farmers to mitigate the effects of environmental degradation and climate change (Ozkan et al., 2022). Climate-smart extension initiatives must be part of a larger set of adaptation strategies and policies for agricultural systems at various scales, including I.C.T. integration (Turyasingura & Chavula, 2022).

#### *8.5.5.1. Climate Smart Agriculture Technologies and Practices*

The technologies and practices supporting the changing climate are listed below:

1. *Seed Smart*: Seeds are the foundation of agriculture, so their resilience is critical to adaptation. In Laos, better rice cultivars and seed systems were identified as the most promising C.S.A. interventions. One of Nepal's most significant C.S.A. efforts were adopting better seeds. In India, stakeholders advocated better seeds as an adaptation tool known as Climate-Smart Seed Management (C.S.S.M.), such as seed treatment with Pseudomonas bacteria, as part of the N.I.C.R.A. project at its adopted community.
2. *Variety Smart*: Extension stakeholders worldwide are screening and introducing location-specific cultivars to mitigate climate change risks and increase productivity. Varieties such as K-9 (drought resistant) and high oil content (53%), K-6 (drought resistant), and

Dharani (drought resistant to groundnut leaf spot disease) contribute significantly to the severe climatic conditions.

3. *Cropping Pattern Smart*: To build climate-smart cropping systems, smallholder farmers should follow practices like using quality seeds adapted to local climate and pests, diversifying crop systems, implementing sustainable mechanization, applying soil and water conservation practices, improving water management, and investing in agricultural knowledge transfer to cope with the effects of changing climate. Fertilizer Smart: Precision fertilizers and remote sensing technology are promising. The Department of Agriculture in India promotes Zero Budget Natural Farming (Z.B.N.F.). The Department of Agriculture's principal extension initiatives for promoting Z.B.N.F. included the formation of a cadre of resource pools known as Master Farmers, Cluster Resource Persons, and Cluster Activists. These extension functionaries attempt to raise fellow farmers' awareness and knowledge of Z.B.N.F.
4. *Water Bright*: In India, both the public and private sectors have developed watersheds to address on-, off---, and non-farm water-related issues. K.V.K. has promoted three check dams (each with a capacity of 0.2 million liters) through the N.I.C.R.A. project to augment water resources and assure the availability of water for irrigation within the project's coverage area.
5. *Knowledge Smart*: I.C.T. advisories and the availability of C.S.A. solutions at reasonable rates can impact farmers' decisions toward adaptation and mitigation. According to F.A.O., more stakeholder cooperation, vital extension outreach, and frequent technical assistance can all help share techniques and innovation. I.C.T. plays an important role in facilitating stakeholders' knowledge.
6. *Community Smart*: Community groups and methods are evolving into cutting-edge extension platforms that increase end users' involvement in C.S.A. design and execution. It is crucial to group decision-making because it convinces and guides farmers to embrace. C.S.A. interventions at the local level.

### *8.5.5.2. Innovative Approaches and Methods for Climate-Smart Agriculture*

1. *Demonstration Effect Approach*: The demonstration effect, sometimes refers to as the progressive farmer strategy, involves

giving credit, inputs, and technical support to a select group of elite farmers who would mature more quickly and thus be better equipped to withstand challenging weather conditions (Zhang, & Hong, 2021; Taylor, M., Bhasme, 2018). This led to the development of a climate-wise practice. The idea was to make up for the absence of extension workers in various regions of Uganda.

2. *Educational and Methods Approaches*: These strategies emphasized farmer education and the employment of extension techniques, including field trips, exposure visits, training at District Farming Institutes, radio and television programs, film screenings, brochures, and posters. As one of the pillars of climate-smart agriculture, these strategies generally led to better farming practices, productivity, and production (Davis, K. (2019; Akumu et al., 2019). (Davis, K. (2019; Akumu, et al., 2019).
3. *Training and Visit Extension and Unified Extension Approaches:* Intermediary staff members must receive continuing education to stay current on issues about climate change, including its effects, implications, and various adaptation and mitigation techniques. They should become more adept at conveying the needs of farmer communities and search for (agro-meteorological) components that need to be addressed.
4. *Climate Farmers Field Schools (FFS):* Through experiential learning, the Farmer Field School (FFS) is a non-formal, participatory extension method that puts farmers' needs first. Consequently, it offers farmers a low setting where they can test out novel agricultural management techniques, talk about and take notes on their observations, and acquire new skills and practical knowledge that will help them make better decisions both individually and as a group (Mfitumukiza et al., 2017).
5. *Climate-Smart Villages (CSVs):* In Uganda, CSVs are developed communities or examples of community-based initiatives that guarantee food security, encourage adaptation, and increase resiliency to climate-related challenges. CSV comprises four components of village development plans, climate-smart technology, local knowledge and institutions, and climate information services (Raghuvanshi et al., 2018; Vincent & Balasubramani, 2021).
6. *Climate Wallpapers:* This helps farmers learn more about climate-smart agricultural practices, which increases productivity. It is used in many parts of Uganda to provide advisory services to farmers in

the form of tables or posters relating to weather forecasts of agricultural activities that the local communities must undertake through display boards.

7. *Using a Public Address System (P.A.S.):* Farmers receive alerts about critical weather conditions or gro-advisory services using this method.
8. *Meetings of the Climate Group*: To connect farmers in different parts of Uganda, including Gulu, Kasese, Kanungu, and Kabale, or to send them climate-related information in a group, these farmers create groups.
9. *Climate Awareness Programmes/Campaigns, Exhibitions*: Extension workers can provide technical inputs by SMS to non-extension players (such as radio or television) who use mass media to raise awareness or provide essential information. For example, a participatory mass extension effort in Malawi has demonstrated the benefits of radio broadcasts on farmers' knowledge of the causes and consequences of climate change, how to make high-quality compost, and how crucial compost is to resilience building. This example also demonstrates how extension strategies that are most effective at increasing public awareness can help mitigate the effects of climate change, promote adaptation, and boost food security (Raghuvanshi et al., 2018; Vincent & Balasubramani, 2021).
10. *Climate Training:* It is crucial to provide education and training to intermediaries and extension staff to ensure they are up to speed on topics such as adaptation and mitigation methods, as well as the implications and repercussions of climate change. Intermediaries for extension should be located closest to farmers and only work at the field level. They should improve their ability to communicate the requirements of farmer communities and look for (agro-meteorological) elements that require attention.
11. *Plant Clinics*: Like human health clinics, plant clinics serve as the primary point of contact for the national extension system and facilitate direct communication between farmers and extension agents on "any problem and any crop." Williams (2015) cites data from a recent study indicating that 82% of farmers in Pakistan, Sri Lanka, Rwanda, Malawi, and Ghana who visited plant clinics said their visits increased crop production (Raghuvanshi et al., 2018).
12. *Contingency Crop Planning*: This document contains suggestions for all significant cultural practices and crop management facets. For

farmers, this type of calendar is beneficial for scheduling irrigation, sowing crops, and taking precautions against damage to plants. Crop planning is participatory in nature since scientists' scientific knowledge and farmers' local knowledge about diverse agricultural methods and crops play a significant role. The agro-meteorological department assists in its preparation (Raghuvanshi et al., 2018).

### 8.5.6. Gender/Women Inclusive Extension Services

Women must participate in agricultural extension services for agriculture-led growth and output to continue. By participating in agricultural extension services, women can improve food security, lower poverty, improve the health and nutrition of their families, open up new career prospects, and increase the effectiveness of extension services. It makes sense to connect women with agricultural extension services for several reasons. Social, cultural, and commercial growth facets can benefit from including gender equality in advising and extension services (Mamun et al., 2017). Women spend most of their time caring for the home, working in agriculture, and performing labour-intensive tasks, yet women encounter several barriers while trying to participate in these activities. Additionally, because of a lack of collective action for empowerment, limited access to resources, prejudices and sociocultural norms, and financial resources, women find themselves at a disadvantage when it comes to using the benefits of extension services. The strategies for addressing the issue of women's low participation in developmental activities include gender-sensitive training programs, community sensitization and awareness campaigns, financial resource access, collective action for empowerment, flexible program design, integration of I.C.T. technologies, and policy advocacy for gender equality (Yadav, & Preethi, 2024).

#### *8.5.6.1. Strategies for Tailoring Extension Services to Meet the Specific Needs of Women*

For gender-inclusive extension services to be relevant and practical, women must actively participate in the assessment process. Women's distinct viewpoints, backgrounds, and objectives become essential to the assessment when participating actively in consultations and cooperative conversations. Here are the notable strategies to fulfill the needs of rural women through extension services:

1. *Flexible Training Schedules*: Since women have distinct roles in the home and agriculture, flexible training programs are designed to accommodate their various schedule constraints. Facilitating weekend or after-hours sessions enhances accessibility and empowers women to engage in extension initiatives without disrupting their daily obligations.
2. *Methods of Practical and Hands-On Learning:* Opportunities for hands-on learning often benefit women. Students are assisted in learning and applying agricultural ideas through engaging seminars, field trips, and practical demonstrations. This method promotes active participation and acknowledges various learning styles.
3. *Accessible Information Formats*: The distribution of information must be adjusted for different literacy levels and preferences. It is ensured that women with various educational backgrounds can effectively acquire and use the information provided during extension services by mixing local languages, using visual aids, and using simple language.
4. *Peer Support Networks Promote Empowerment*: Developing peer support networks facilitates women's sharing of experiences, challenges, and victories. Peer mentorship creates a venue for knowledge exchange and emotional support, which fosters a sense of community. Encouraging women via these networks increases their self-esteem and reinforces the efficacy of outreach programs.
5. *Customized Training Modules*: Training materials that specifically address the needs and interests of women are essential, which means that information must be changed to highlight certain crops or behaviors that women frequently engage in. Gender-specific examples and case studies must also be included to improve application and relevance.
6. *Adoption of Inclusive Technology*: Access to extension services can be enhanced through technology, such as internet platforms and smartphone apps. By providing training materials in digital formats, barriers like time constraints or geographic distance can be overcome and women can access knowledge at their own pace.
7. *Culturally Sensitive Approaches:* Understanding and upholding cultural norms is crucial. By engaging with local communities to understand cultural sensitivity, extension services are guaranteed to align with community values. This approach recognizes and

considers cultural considerations in the design of the curriculum, increasing acceptability and engagement.

8. *Decision-Making Empowerment:* Women are strengthened when they are actively involved in agricultural decision-making processes. This method ensures that women are included in the decision-making process regarding resource management, practice implementation, and crop choice. Participating in the decision-making process helps to promote more impactful and sustainable farming practices in addition to recognizing their skills.

## 8.5.7. Extension Efforts in Promoting Youth Engagement and Employment in Agriculture

It's a prevalent belief that young people represent the future of humanity. One of the most critical factors in attaining sustainable development is the youth, especially in developing nations where most of them live in rural areas. In every part of the world, the unemployment rate for young people is three times higher than that of adults, and young women make up the bulk of those without jobs. Along with significant obstacles to land, credit, and other productive resources needed to start their own businesses, youth also frequently lack the right to union participation in producers' associations or labor unions.

The youth of today are in charge of transforming the food and agricultural systems. They are adjusting to the impacts of climate change and the environment, which will probably deepen and accelerate throughout their lifetimes F.A.O. (2021). Before the COVID-19 pandemic, their unemployment rate was around three times higher than that of adults. The current crisis threatens to worsen already-existing disparities both within and across nations. I.L.O.'s 2024 report on global employment and social network trends revealed that, while slightly better than in 2022, the worldwide unemployment rate in 2023 was still 5.1%, a cause for concern. The primary goal of extension at the time was adult education, and agricultural extension programs were primarily created with this widespread rural illiteracy in mind, concentrating on treatments meant to increase the functional literacy of rural adults.

### *8.5.7.1. Initiatives to Enhance Capacities of Rural Youth for Extension Services*

1. *A.R.Y.A. (Attracting and Retaining Youth in Agriculture):* In 2015-16, the Indian Council of Agriculture Research in New Delhi conducted this project through its agricultural extension division. This effort aims to attract, empower, and include rural adolescents in various agricultural and related sector entrepreneurial ventures. Project A.R.Y.A. is carried out in 25 states via Krishi Vigyan Kendra (K.V.K.s). The project involves selecting one district from each state. Technical collaborators for the project include I.C.A.R. Institutes and Agricultural Universities. 200-300 rural teenagers (in one district) are identified and given entrepreneurship and skill development training so that they can start micro-enterprises in fields such as apiary, mushroom, seed processing, dairy, goats, poultry, carp hatchery, vermicomposting, and so on. One or two enterprise units will be established at Krishi K.V.K.ss Kendras (K.V.K.s) to act as entrepreneurial training facilities for farmers (Chander, 2013).
2. *National Rural Livelihood Mission (N.R.L.M.):* NRLM strives to create chances for both wage employment and skill development for rural youth who lack skills in many agricultural production and processing areas.
3. *Agricultural Skill Council of India (ASCI):* The National Skill Development Corporation (N.S.D.C.) has proposed the Agricultural Skills Council of India (ASCI), which might be a perfect institution to train rural youth, in light of the necessity to skill the labor force in the agricultural industry. Through its training partners, the ASCI plans to teach, certify, and accredit 56.5 million workforces over ten years, including wage workers, farmers, entrepreneurs, and extension workers.
4. *ASEAN-India Farmers Exchange:* To achieve more sustainable agriculture development in the region, the Ministers of the Association of Southeast Asian Nations, or ASEAN, expressed concern over the low number of young farmers involved in the agriculture sector and agreed on encouraging innovation and entrepreneurship among young farmers. The first exchange visit occurred in November 2012 in Malaysia in connection with the ASEAN Farmers' Week and the Malaysian Agriculture, Horticulture and Agro-tourism (MAHA) International. A delegation of farmers

from nine ASEAN member states participated in the second exchange visit, which took place in India from December 19 to 30, 2012.

5. *Young Farmers' Associations:* The U.S.'s rural youth programming was sparked by young people's passion for new technology, and as a result, 4-H clubs were founded in 1914. Promoting and modernizing agriculture is the focus of the recently established "The Young Farmers Association (Y.F.A.), Punjab" in India.
6. *Strategies for Enhancing Youth Participation in Agriculture:* To inspire youths to pursue agricultural occupations, we need a multifaceted approach to addressing their problems while emphasizing the benefits available (Choudhary et al., 2023).
7. *Raising Awareness:* Organise job fairs and awareness campaigns to showcase the range of agricultural career options and the industry's potential for expansion and innovation.
8. *Providing Resources and Support:* Implementing programs and policies that enable young agri-preneurs with access to land, finance, and other resources, such as the Agri-Clinics and Agri-Business Centres Scheme (A.C.A.B.C.) and the Pradhan Mantri Kisan Samman Nidhi.
9. *Investing in Education:* Redesigning agricultural training and education courses to incorporate new technology, methodologies, and agribusiness management skills that appeal to the interests of young people.
10. *Leveraging Technology:* Advocate for new agricultural technologies and digital tools to make the industry more appealing and efficient to younger generations.

## Conclusion

Extension-dependent services' future depends on their ability to embrace diversity, creativity, and inclusivity. By taking a holistic and participative approach, extension services can continue to be positive change agents, advancing sustainable agricultural growth and enhancing livelihoods for future generations. The function of agricultural extension in the coming decade will be very different from what it was ten years ago or even now. Its importance as a facilitator of the agricultural knowledge system will only grow as more private-sector entities engage in extension. It is conceivable that input-related extensions will transfer to the private sector. Recognizing the critical nexus between agriculture and nutrition, extension services increasingly focus on promoting diverse and nutrient-rich food production. By emphasizing the nutritional value of crops and advocating for dietary diversity, extension efforts contribute to improved food security and public health outcomes. Building resilience to climate change is imperative for the sustainability of agriculture. Extension services promote climate-smart practices, resilient crop varieties, and adaptive strategies to mitigate risks and enhance productivity in the face of climate variability. Recognizing the significant role of women in agriculture, extension efforts prioritize gender equality and women's empowerment. Extension services should provide more opportunities for skill development, entrepreneurship training, and access to resources to attract and retain young people in agriculture. For more inclusive extension services globally, government intervention should pay attention to extension services by including the multi-facets of extension in the advanced era.

## References

Akinnagbe, O. M., & Ajayi, A. R. (2010). Challenges of farmer-led extension approaches in Nigeria. *World Journal of Agricultural Sciences*, *6*(4), 353-359.

Akumu, G., Atukwase, A., Tibagonzeka, J. E., Apil, J., Kato, J. S., Kiyimba, F., & Muyonga, J. H. (2019). The influence of extension approaches on uptake of post-harvest technologies among maize farmers in Uganda. *African Journal of Rural Development*, 3(4), 955-971.

Arumugam, U., & Manida, M. (2023). Agripreneurship for Sustainable Economic Development in India. *ComFin Research*, *11*(4), 15-23.

Ashraf, I., Muhammad, S., Mahmood, K., Idrees, M., & Shah, N. (2009). Strength and weaknesses of extension system as perceived extension field staff. *Sarhad J. Agric*, *25*(1), 131-134.

Babadoost, M., (2018). Agricultural Extension in the United States of America: History and Problem- Solving, in Jayawardena, L.N.A.C., Suryamani, M., and Sivayoganathan, C. (Eds) *Proceedings, International Conference in Transforming Agriculture Extension Systems: Towards Achieving Relevant Sustainable Goals for Global Impact.* Faculty of Agriculture, University of Peradeniya, Peradeniya, Sri Lanka.

Bachano, T., & Jimma, E. (2019). Paradigm Shift in Agricultural Extension. *Rural Planning and Development,* https://doi.org/10.1223.

Barber, J., Mangnus, E., & Bitzer, V. (2018). Harnessing I.C.T. for agricultural extension. *Europe*, *91*(120.6), 31-35.

Behera, B., Mallick, B., Ayushree, B. A., & Singh, P. (2024). From Tradition to Transformation: The History of Agricultural Extension, *Publish My Work publication.* ISBN: 978-81-96458-84-3.

Bull, N. H., Cote, L. S., Warner, P. D., & McKinnie, M. (2004). Is extension relevant for the 21st century? *The Journal of Extension*, *42*(6), 3.

Chander, M. (2013). Youth: Potential target for agricultural extension. *Agricultural Extension in South Asia.*

Choudhary, P., Jaitawat, V. S., Singh, S. S., & Saikanth, D. R. K. (2023). Empowering Women and Youth in Agriculture Through Extension. *Systems in Agri-Horti.* ISBN: 978-81-967311-7-5.

Davis, K. (2019). The complex processes of agricultural education and extension. *The Journal of Agricultural Education and Extension*, *25*(3), 193-194.

Davis, K., Dolly, D., Lamm, A. J., Lamm, K. W. (2018). The future of extension: A network emergence perspective from the case of the Global Forum for Rural Advisory Services." *Journal of International Agricultural and Extension Education*, 25(4), 40–51. https://doi.org/10.5191/jiaee.2018.25403.

Ezeh Ann, N. (2013). Extension agents access and utilization of information and communication technology (I.C.T.) in extension service delivery in Southeast Nigeria. *Journal of Agricultural Extension and Rural Development*, *5*(11), 266-276.

F.A.O. (2021). Promoting youth engagement and employment in agriculture and food systems. Committee on World Food Security, http://www.fao.org/3/cb5463en/cb5463en.pdf.

F.A.O. Committee on Agriculture, Twenty-fifth Session Rome, 26 - 30 September 2016 Second International Conference on Nutrition (ICN2) Follow-up: Nutrition-related Implications for Agriculture and Livestock Development.

Franzel, S., Degrande, A. Kiptot, E. Kundhlande, G. Tsafack, S. and Simpson, B. In press. Does farmer-to-farmer extension increase women's participation and access to advisory services? Lessons from Kenya, Cameroon and Malawi. *Journal of International Agricultural and Extension Education,* in press.

Franzel, S., Degrande, A., Kiptot, E., Kirui, J., Kugonza, J., Preissing, J., & Simpson, B. (2015). Farmer-to-farmer extension. *What works in rural advisory services? Global good practice notes*, 53-56.

Gangwar R. & Jadoun, R. S. (2023). The Government Interventions in Strengthening Agripreneurship in India: A Comprehensive Study. *International Journal of Arts & Education Research,* 12(5), 85-90.

Garforth, C. (2010). Adapting to new challenges: extension theory and practice for the 21st century. In: International Conference on Agricultural Extension (Agrex 10), 26-28 October 2010, Universiti Putra Malaysia, pp. 1-11. (Keynote Paper).

Garforth, G. E., Jones, C., Jones, G., & Garforth, C. (1997). The history, development, and future of agricultural extension. *Improving agricultural extension: a refere F.A.O manual.* Roma: F.A.O.

Jha, S., Kashyap, S. K., Ansari, M. A., Kamswari, V. L. V., Singh, S., & Goswamy, P. (2021). Attitude of Farm Women Towards I.C.T. tools-based extension services. *Indian Res. J. Ext. Edu,* 21(1), 96-98.

Jones, G. E. (1994). Agricultural advisory work in England and Wales: The beginnings. *Agricultural Progress, 69,* 55-69*Journal d'Agriculture Pratique* (1874). 38th year, vol. 2.

Kafura, R. A., Afrad, M. S. I., & Chakraborty, F. A. P. D. B. (2016). Use of I.C.T. as extension tool by the farmers of Gazipur district in Bangladesh. *Indian research journal of extension education*, *16*(2), 1-5.

Kassa, B., & Temesgen, D. (2011). Perspective of Agricultural Extension (RDAE 511). College of Agriculture and Environmental Sciences, Department of Rural Development and Agricultural Extension. November 2011, Haramaya University. (Module material).

Kumar, V., Lehri, S., Sharma, A. K., Meena, P. D., & Kumar, A. (2016). Image based rapeseed-mustard disease expert system: An effective extension tool. *Indian Research Journal of Extension Education*, *8*(3), 10-13.

Kumari, Anjali Pandey, Bhanu Prakash Mishra (Eds), pp. 121-131, Elite Publishing House, New Delhi, ISBN: 978-93-58990-52-2.

Kumari, Veenita., Junuthula, Shirisha. 2023. *Strengthening Capacities of Extension Advisory Service (E.A.S.) Providers in making Food Systems Gender and Nutrition-sensitive: MANAGE Experiences. Presentation.* Presented at the CGIAR GENDER Conference' From Research to Impact: Towards just and resilient agri-food systems', New Delhi, India, 9-12 October 2023. National Institute of Agricultural Extension Management, India.

Lakshmi, K. B., & Naidu, D. C. (2022). *Extension strategies for agripreneurship promotion in tribal and rural areas in Srikakulam District of Andhra Pradesh.*

Lemmon, H. (1986). C.O.M.A.X.: An expert system for cotton crop management. *Science*, *233*(4759), 29-33.

Mahmoud, M., Rafea, M., & Rafea, A. (2008, November). Using expert systems technology to increase agriculture production and water conservation. In 2008 *Third International Conference on Digital Information Management* (pp. 1-7). IEEE.

Maier-Bode, F. (1910). *Die Organization und die Erfolge des landwirtschaftlichen Wanderunterrichts im Königreich Bayern.* Landsberg am Lech: Georg Verza.

Mallick, B., Anshuman, J., Behera, B. and Rai, R. (2023). *Extension Systems in India. In: Advances in Agricultural Extension* Volume 1.

Mallick, B., Anshuman, J., Behera, B. and Rai, R. (2023). Extension Systems in India. In: *Advances in Agricultural Extension* Volume 1, Somdutt Tripathi, Brijesh Kumar Gupta, Jagriti Kumari, Anjali Pandey, Bhanu Prakash Mishra (Eds), pp. 121-131, Elite Publishing House, New Delhi, ISBN: 978-93-58990-52-2.

Mamun-Ur-Rashid, M., Kamruzzaman, M., & Mustafa, E. (2017). Women participation in agricultural extension services in Bangladesh: Current status, prospects and challenges. *Bangladesh Journal of Extension Education* ISSN, 1011, 3916.

Meena, M. S., Kale, R. B., & Singh, S. K. (2018, April). LEAD Farmer based Farmer to Farmer Extension Model: SWOT Analysis and Sustainability Issues in Indian Context. In Lead paper published in Souvenir of National Seminar on Extension Strategies for Doubling the Farmers' Income for Livelihood Security organised by Society of Extension Education, Gujarat and Anand Agricultural University, Anand, Gujarat, India, during (pp. 26-27).

Meena, M. S., Kale, R. B., Singh, S. K., & Gupta, S. (2016). Farmer-to-Farmer Extension Model, *Sustainability& Scalability in Indian Perspective.*

Mfitumukiza, D., Barasa, B., Nankya, A. M., Dorothy, N., Owasa, A. H., Siraj, B., & Gerald, K. (2017). Assessing the farmer field schools diffusion of knowledge and adaptation to climate change by smallholder farmers in Kiboga District, Uganda. *Journal of Agricultural Extension and Rural Development*, *9*(5), 74-83.

Mishra S., Akankasha, Deep V., (2014). Expert Systems in Agriculture: An Overview, *International Journal of Science Technology & Engineering, 1(5),* 45-49.

Mohapatra, T., Goyal, R., & Goyal, C. (2021). *Role of internet of things in sustainable agriculture in India.*

Mook, B. T. (1982). *The world of the Indian field administrator.* New Delhi: Vikas Publishing House.

Murali, S., & Thomas, K. V. (2022). Motivational Factors for Entering into Agri-preneurship: A Study among Successful Agri-preneurs of Kerala. *Journal of Extension Education*, *34*(1), 6784-6787.

Nakano, Y., Tsusaka, T. W., Aida, T., & Pede, V. O. (2018). Is farmer-to-farmer extension effective? The impact of training on technology adoption and rice farming productivity in Tanzania. *World Development*, 105, 336-351.

Nedumaran, S., & Ravi, N. (2019). Agriculture extension system in India: A meta-analysis. *Research Journal of Agricultural Sciences*, *10*(3), 473-479.

Ochieng, W., Silvert, C. J., & Diaz, J. (2022). Exploring the impacts of lead farmer selection on community social learning: The case of farmer-to-farmer model: A review of literature. *Journal of International Agricultural and Extension Education*, 29(3), 7-31. doi: https://doi.org/10.4148/2831-5960.1022.

Ozkan, B., Dube, A. K., & Reed, M. R. (2022). Role of agricultural extension in building climate resilience of food security in Ethiopia. In *Food Security and Climate-Smart Food Systems: Building Resilience for the Global South* (pp. 219-239). Cham: Springer International Publishing.

Paroda, D.R. (2010). Workshop on ICT/ICM for National Agricultural Research Information Systems in the Asia-Pacific Region. Jointly Organized by A.P.A.A.F.A.OG.F.A.R.FA.I. T G.F.A.R. and A.I.T. 14-16 September 2010. p: 1.

Raghuvanshi, R., Raj, S., & Bhattacharjee, S. (2018). *Climate smart agriculture and advisory services: approaches and implications for future*. Discussion Paper 1, Center for Agricultural Extension Innovations, Reforms and Agripreneurship (C.A.E.I.R.A.), Rajendranagar, Hyderabad.

Rani, P. M. N., Rajesh, T., & Saravanan, R. (2011). Expert systems in agriculture: A review. *Journal of Computer Science and Applications*, *3*(1), 59-71.

Ray, P., Panigrahi, R. S., & Shasani, S. (2022). Determinants of skill levels of farm youth with regard to agripreneurship: A multinomial regression approach. *Indian Journal of Extension Education*, *58*(1), 58-62.

Rivera, W. M., & Alex, G. (2004). Extension system reform and the challenges ahead. *The Journal of Agricultural Education and Extension*, *10*(1), 23–36. https://doi.org/10.1080/13892240485300051.

Sahoo, S. K., Gaurav, S., & Meeyo H. (2023). Feeding The Future: Nutrition-Sensitiv Extension- A Key Ingredient for Global Food Security. *Agriallis: Science for Agriculture and Allied Sector: A Monthly Magazine*, 5(11).

Sangeetha, V., Singh, P., Mahra, G. S., Sarkar, S., Satyapriya, S., & Upadhyay, S. (2016). Agri-Nutri Smart Villages A leap towards ensuring nutritional security in India. *Indian Farming*, *66*(1).

Sarma, S. K., Singh, K. R., and Singh, A., 2010," An Expert System for diagnosis of diseases in Rice Plant," *International Journal of Artificial Intelligence* 1 (1): 1-6.

Scarbourough, V., S. Killough, D. A. Johnson and J. Farrington (eds.). 1997. *Farmer-led Extension: Concepts and Practices.* London: Intermediate Technology Publications.

Shinn, G. C., Wingenbach, G. J., Lindner, J. R., Briers, G. E., & Baker, M. (2009). Redefining agricultural and Extension education as a field of study: Consensus of fifteen engaged international scholars. *Journal of International Agricultural and Extension Education,* 16(1), 73-88. https://www.aiaee.org/attachments/101_Shinn-Redefining-Vol-16.1%20e-6.pdf.

Singh, N. K., Sunitha, N. H., Tripathi, G., Saikanth, D. R. K., Sharma A., Jose A. E., & Mary, M. V. Karuna Jeba (2023). Impact of Digital Technologies in Agricultural Extension, *Asian Journal of Agricultural Extension, Economics & Sociology,* 41(9), 963-970.

Smith, E. G., & Smith, R. D. (2007). Will Extension be Relevant In the 21st Century? Perspectives on 21st Century Agriculture: A Tribute to Walter J. Armbruster, 56.

Srinivasan, R., Engel, B. A., & Paudyal, G. N. (1991). Expert system for irrigation management (E.S.I.M.). *Agricultural Systems*, *36*(3), 297-314.

Suresh, A., Sajesh, V. K., Padaria, R. N., & Mohanty, A. K. (2022). Reinventing Agricultural Extension System in India the Road Ahead.

Taylor, M., & Bhasme, S. (2018). Model farmers, extension networks and the politics of agricultural knowledge transfer. *Journal of Rural Studies*, *64*, 1-10.

The triple burden of malnutrition. *Nat Food* **4**, 925 (2023). https://doi.org/10.1038/s43016-023-00886-8.

Turyasingura, B., & Chavula, P. (2022). Climate-smart agricultural extension service innovation approaches in Uganda.

Tzounis, A., Katsoulas, N., Bartzanas, T., & Kittas, C. (2017). Internet of Things in agriculture, recent advances and future challenges. *Biosystems engineering*, *164*, 31-48.

Uccello, E., Kauffmann, D., Calo, M., & Streissel, M. (2017). Nutrition-sensitive agriculture and food systems in practice: options for intervention.

United Nations Children's Fund (UNICEF), World Health Organization (WHO), International Bank for Reconstruction and Development/The World Bank. Levels and trends in child malnutrition: UNICEF/WHO/World Bank Group Joint Child Malnutrition Estimates: Key findings of the 2023 edition. New York: UNICEF and WHO; 2023. CC BY-NC-SA 3.0 IGO.

Vincent, A., & Balasubramani, N. (2021). Climate-smart agriculture (C.S.A.) and extension advisory servic E.A.S.EAS) stakeholders' prioritisation: a case study of Anantapur district, Andhra Pradesh, India. *Journal of Water and Climate change*, *12*(8), 3915-3931.

Vinod Kumar, Sushma Lehri, Ashok Kumar Sharma, P. D. Meena and Arvind Kumar, 2008, "Image Based Rapeseed-Mustard Disease Expert System: An Effective Extension Tool," *Indian Res. J. Ext. Edu.* 8 (2&3).

Weinand, J. (2002). Farmer-to-farmer extension: Opportunities and constraints of reaching poor farmers in southern Malawi. *Unpublished master's Thesis. University of Hohenheim, Germany.*

Wijeratne, M., & De Silva, W. N. (2024). Babadoost, M., (2018). Bachano, T., & Jimma, E. (2019). Sangeetha, V., Singh, P., Satyapriya, Lenin, V., Paul, S., Barua, S., Muralikrishnan, L., and Bishnoi, S. (2018). *An Agri-Nutri (A2N) Smart Village Model for Food and Nutrition Security*. Division of Agricultural Extension, ICAR-Indian Agricultural Research Institute, New Delhi- 110 012.

Wijeratne, M., & De Silva, W. N. (2024). *The Paradigm Shift of Agricultural Extension from Technology Transfer Towards Participatory Approaches.*

Xu, J., Gu, B., & Tian, G. (2022). Review of agricultural IoT technology. *Artificial Intelligence in Agriculture*, *6*, 10-22.

Yadav, M. K., & Preethi, M. (2024). Gender Inclusive Extension Services: Empowering Women in Agriculture. *Stella International Publication,* ISSN-9788196723842.

Zhang, P., & Hong, J. (2021). Education, Local Financial Investment and Rural Financial Repression: Learn from the U.S.'s Demonstration Effect on China. *CONVERTER*, 2021(7), 27-36.

## Chapter 9

# Farm to Fork: Building Sustainable Food Value Chains Through Extension

**Adupa Shanmuka**[1,*]
**T. N. S. S. Srivani**[2]
**and Veesam Haripriya**[3]
[1]ICAR-Directorate of Groundnut Research, Junagadh, Gujarat, India
[2]Department of Agricultural Extension Education, Junagadh Agricultural University, Junagadh, Gujarat, India
[3]Department of Agricultural Extension Education, ICAR-Indian Agricultural Research Institute, New Delhi, India

## Abstract

In agriculture, the term "value chain" refers to the broad range of actions and parties involved in moving agricultural products from production to consumption, emphasizing adding value at each stage. In contrast to a conventional supply chain, a value chain prioritizes customer-focused approaches. Developing sustainable agriculture value chains leads to high income, employment, foreign exchange earnings, and a new method to combat food security challenges. Food value chains face problems at every stage, from producer to processor to consumer, in different ways. Hence, a multidimensionally focused analysis and the ability to address the difficulties help develop more sustainable value chains. To address the issues of every stakeholder, we must provide an innovative solution with new technological advancements. An integrated approach to value chain management is required to overcome all the problems and adapt to the latest marketing realities in food value chains. Big Data, IoT, data

* Corresponding Author's Email: shanu23197@gmail.com

In: Global Horizons in Agricultural Extension Education
Editors: R. K. Doharey, Aman Verma, Satyapriya et al.
ISBN: 979-8-89530-446-4

analytics, Blockchain technology, artificial intelligence, cloud computing, and robotics, which are interoperable technology-enhanced platforms in the form of service providers, are providing solutions to supply chain management problems. By bridging gaps in the agro-value chain, startups can offer farmers and consumers efficient goods, technology, and services. Agricultural Extension services play a multifaceted role in this integration. Agricultural Extension services contribute to capacity building to foster entrepreneurship through incubation and technology backstopping for new startups. Startups developing in the line of providing solutions to problems may be encouraged, and new Agripreneurship development activities can help promote the startup ecosystem in the country. Promoting farmer-producer organizations can help create more market access for the farming community and enhance their quality standards by increasing the technology adoption rate. FPOs lead to risk mitigation for farmers and increase their access to finance. Agricultural extension plays a crucial role in mobilizing farmers to form producer organizations and ensuring the sustainability of farm-producer organizations by providing the knowledge, resources, and skills needed to implement sustainable and profitable farming practices. This support improves the long-term resilience and success of FPOs. These strategies can be combined to create a sustainable food value chain that supports social responsibility, environmental preservation, and commercial viability from farm to fork.

**Keywords:** value chain, sustainable agriculture, multidimensional, FPO, food security

## 9.1. Introduction

India's economy revolves mainly around agriculture, the foundation of the nation's socioeconomic growth. About two-thirds of the people depend on it, contributing roughly 19% of the GDP. Because of its backward and forward links, the success of agriculture dramatically influences the growth of other sectors and the economy. In addition to providing food security and a means of subsistence for a substantial portion of India's population, it holds particular importance for the underprivileged, poor, and marginalized groups. Despite favorable policy conditions and robust demand, India's agriculture sector faces many challenges. These include, among other things, smaller holding sizes, inadequate infrastructure for primary and secondary processing, a convoluted supply chain with numerous levels and intermediaries, and restricted last-mile service offerings. Value chains refer to the structured connections between groups of producers, traders, processors, and service providers, including nongovernmental organizations (NGOs), that collaborate to increase productivity and add value to their operations. The entire process of producing, processing, distributing, and consuming agricultural commodities—from farm to fork—is called an agri-commodity value chain. A supply chain and a value chain are not the same thing. Value chains comprise linkages that provide value to the consumer—the procedures involved in transferring and transforming goods into products from manufacturers to end users. A supply chain is about logistics, whereas a value chain is about creating value for the customer.

Many socioeconomic issues, including the prevalence of small and marginal farmers, fragmented supply chains, a lack of scale economies, low processing and value addition levels, limited marketing facilities, etc., impact the nation's agricultural supply chain system. Various supply chain stakeholders have varying sets of issues that need to be investigated, including:

### 9.1.1. Producer-Level Issues

High credits trap people in debt or poverty, insufficient infrastructure and limited access to superior inputs such as loans and technologies, inadequate extension services, production management that is ineffective and insufficient, low level of output unrelated to demand production, poor investment returns, high commission fees, high cost of transporting, lack of farm gate marketing options, distressed sales, delinking of prices from the market, food crops with low marketable surpluses and no mechanism for aggregation, lack of adequate warehousing facilities for perishable agricultural produce, such as cold chain, cold storage, etc., inadequate production and market risks are addressed through risk mitigation actions, poor quality as a result of improper postharvest management.

### 9.1.2. At the Level of Market Intermediaries

Not a player who adds value, charges farmers large commission fees, unorganized trading's dominance, acts restricting APMC (though liberalized now), losses, both qualitative and quantitative, unscrupulous profit-making, multiple levels in the marketing channel are raising prices, which isn't in the producer's or consumer's best interests, inadequate produce handling, sorting, and grading infrastructure and a lack of grading, no connections, no price transparency, long delays between the producer and the retailer.

### 9.1.3. Main Concerns with Logistics

Internet access in rural areas, open vehicles (especially for transporting perishable goods), and the lack of an integrated cold chain remain significant challenges. Progress has been slow, and high transportation costs exacerbate the situation.

#### *9.1.3.1. Processor/User Industry*

In a long or fragmented supply chain, too many intermediaries add little or no value. In the absence of the producer or middleman adhering to grading, there are no controls over the quality of the raw materials sourced, traceability and product standards concepts have not yet gained widespread adoption, laws governing contract farming are rarely upheld, issues with increased procurement costs and variable raw material supply, food safety and hygiene concerns, mainly because there is a sizable "unorganized" processing sector, processing speed, poor quality, poor profits, reduced capacity usage.

Based on the challenges mentioned above, there is high time for a shift to a more integrated and digitalized supply chain management where all the shortcomings can be addressed. , guarantee sustained agricultural output, satisfy consumer needs, tackle global sustainability issues, encourage stakeholder accountability, and adjust to the effects of climate change, agri-commodity value chains must be made sustainable. Agri-tech, being the solution provider to many dimensions of agriculture, has paved the way for solutions to the problem in supply chain management. Here, technologies like big data, IoT, data analytics, Blockchain technology, Artificial intelligence, robotics, and cloud computing are making more accessible and interoperable technology-enriched platforms in the form of service providers. "We must safeguard the environment to solve supply chain management problems." As a result, the country has several opportunities for agricultural entrepreneurs. The conversion of agriculture into agribusiness is one of the primary strategies business-minded farmers employ to achieve thriving agriculture. The growth of agribusiness is a crucial component of agricultural value chains. The term "agribusiness" refers to the connections between farm inputs, product output, and all other commodities and services required to produce and market an agricultural commodity. It is the off-farm link in agrifood value chains. It connects the farm subsector to customers and supplies inputs to it. Agritech-based agribusiness can offer pertinent and cutting-edge answers to various agricultural value chain problems (Shruti, 2018).

## 9.2. Agri Startup Business Models Are of Three Types

Figure 9.1 illustrates the role of Agri Startups in the Food Value Chain and they are of three types:

### 9.2.1. Margin-Based Model

Segments such as "market linkage – farm inputs" and "supply chain tech and output market linkage" operate through this model where the Agritech player earns margin by creating marketplace linkages at the input or output side and offering fulfillment services.

### 9.2.2. Subscription-Based Model

Agri-tech players in "precision agriculture and farm management," "quality management and traceability" segments offer a mix of hardware, software, and services-based solutions across the year and levy monthly or annual subscription charges from their customers.

### 9.2.3. Transaction-Based Model

Players in the "financial services" segment deal follow a transaction-based pricing model based on the number of loans or insurance policies served.

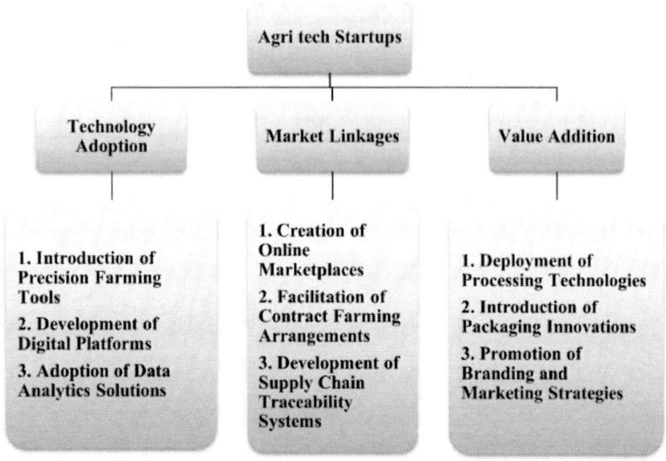


Source: Author Compilation.

**Figure 9.1.** Role of Agri Startups in the Food Value Chain.

## 9.3. Agri-Tech Startups Are Providing Their Services in Various Areas (NASSCOM, 2019)

### 9.3.1. Better Access to Inputs

Taking quality inputs to the farmers for better productivity. Improved availability of agricultural inputs at farmers' doorsteps aids in their understanding of the appropriate input product to boost yield and productivity. Interventions such as analytics, e-commerce, and mobile apps are used. As illustrated in Table 9.1 startups enabling online procurement of quality inputs.

**Table 9.1.** Startups enabling online procuring of quality inputs

| Startup | Overview | City | Year |
|---|---|---|---|
| Agrostar (Soonicorn) | Online service for agricultural inputs, information, and guidance | Pune | 2008 |
| Khethinext | Agricultural inputs can be purchased through a mobile app, which also offers information | Hyderabad | 2017 |
| Gramophone | An app-based platform that offers farmers information and agriculture input items | Indore | 2016 |

Source: Nuthalapati (2020).

### 9.3.2. Digital Agriculture

As Table 9.2 presents startups offering efficient mechanization and irrigation. Driving transparency, traceability, and real-time access to information are the primary targets of digital agriculture. Precision and Digital Startups based on agriculture provide cutting-edge technological solutions to boost crop productivity and farming efficiency. They use technology interventions such as drones, IoT Sensors, blockchain, image sensing, analytics, machine learning.

**Table 9.2.** Startups offering efficient mechanization, irrigation

| Startup | Overview | City | Year |
|---|---|---|---|
| FarMart | Web and mobile-based applications for renting farm equipment | Gurgaon | 2015 |
| EM3 Agri Services | Provider of farming services to the farming communities | Noida | 2013 |
| RAVGO | Digital farm and construction Equipment rentals marketplace | Gurgaon | 2015 |

Source: Nuthalapati, (2022).

### 9.3.3. Farming as a Service

As illustrated in Table 9.3, Farming as a Service provides farmers with accessible technological solutions for practical farming by reducing fixed costs to variable costs. Buying agricultural equipment requires enormous amounts of money, making it difficult for the farmer. A cyclic business model, as farmers need equipment regularly. 75% of farmers own less than 1 hectare of land, making it challenging to own equipment.

**Table 9.3.** Startup offering farming as a service

| Farming as a Service | | | |
|---|---|---|---|
| Startup | Overview | Year | City |
| Farmizen | Develops and operates the digital application for community-supported farming | 2017 | Bangalore |
| Triton Foodworks | Integrated business for soil-less cultivation of fruits & vegetables and supply of produce | 2014 | Delhi |
| Agro20 | Manufacturer and supplier of indoor hydroponics system | 2017 | Delhi |

Source: Nuthalapati, (2022).

### 9.3.4. Financing

Farmers in India have difficulty obtaining funding. Agri-tech-based financing startups make it easier for these marginalized communities of farmers to get loans easily which is very well represented in Table no 9.4. They use technology interventions such as app-based platforms, analytics and data science.

**Table 9.4.** Startup offering financing

| Financing | | | |
|---|---|---|---|
| Startup | Overview | Year | City |
| SG Agtech Innovations | Online platform for providing digital & financial solutions to farmers | 2018 | Chennai |
| SafalFasal | Online market- place for agricultural products | 2019 | Mumbai |
| Jai Kisan | Online supply chain platform for farmers | 2017 | Mumbai |

Source: Nuthalapati, (2022).

### 9.3.5. Improving Supply Chain

One of the easiest ways to get farmers' products to end users is through Market Linkage. This digital platform connects agricultural production with the customer. Technology intervention used was an app-based platform, image sensing for quality grading and IoT-based storage monitoring. It helps in developing an efficient supply chain. Directly taking farm output to the consumer, tremendous customer traction can be seen in this business area, it received the highest funding among all other business areas and helped reduce postharvest losses from 4% to 16% of total output Table 9.5 presents startups helping in postharvest management.

**Table 9.5.** Startup Helping in Postharvest Management

| Startups Helping in Postharvest Management | | | |
|---|---|---|---|
| Startup | Overview | Year | City |
| Ecozen Solutions | Manufactures and supplies solar-powered irrigation pump controllers | 2009 | Pune |
| SaptaKrishi (Sabjikothi) | Provider of a micro-climate storage solution for farmers | 2018 | Kanpur |
| New Leaf Dynamic Technologies | Off-grid refrigeration | 2012 | Delhi |

Source: Nuthalapati, (2022).

**Table 9.6.** Startups for Quality Monitoring, Maintenance, and Predictions of Crop Health and Output

| Startups for Quality Monitoring, Maintenance, and Predictions of Crop Health and Output | | | |
|---|---|---|---|
| Startup | Overview | Year | City |
| CropIn | Provider of saas-based farming solutions to agribusinesses | 2010 | Bangalore |
| Intello Labs | Image recognition-based solutions for multiple industries | 2016 | Bangalore |
| FarmERP | Software suite for control over farm operations and traceability | 2005 | Pune |

Source: Nuthalapati, (2022).

## 9.4. The Role of Extension in Mobilizing Agri Startups to Promote Food Value Chains

Several types of startups have come up in the last decade that are filling the gap in the food value chains in infrastructure deficit regions of the country and introducing innovations which is very well illustrated in Table 9.6. Most of these startups operate in tandem with various related companies downstream, like supermarkets, retailers, and hoteliers; in the midstream with the processors, wholesalers, and logistic firms; and in the upstream with the input companies. The entry of startups has accelerated flows amongst food chain actors concerning making and diffusing innovations to the end users. These service providers offer various services to improve productivity, efficiency, and profitability across multiple stages of the agricultural value chain. Remember that small enterprises are delicate, especially initially. Support in the form of suitable financial and non-financial services is therefore required. Offering various business development services—particularly nonfinancial services—becomes essential for entry, survival, productivity, competitiveness, and expansion for small farmers and companies. There are several types of startups that have come up in the last decade that are filling the gap in the food value chains in infrastructure deficit regions of the country and introducing innovations. Most of these startups operate in tandem with various other related companies downstream, like supermarkets, retailers, and hoteliers; in the midstream with the processors, wholesalers, and logistic firms; and, in the upstream with the input companies and so on. The entry of startups has accelerated flows amongst food chain actors concerning making and diffusing innovations to the end users. These service providers offer a range of services to improve productivity, efficiency, and profitability across various stages of the agricultural value chain. It is important to remember that small enterprises are quite delicate, especially in the beginning. Support in the form of suitable financial and non-financial services is therefore required. Offering various Business Development Services—particularly nonfinancial services—becomes essential for small farmers' and companies' entry, survival, productivity, competitiveness, and expansion (Sharma, 2022).

### 9.4.1. Business Development Services in the Agricultural Value Chain

Individuals and institutions that assist in the manufacturing, distribution, and marketing of products but do not themselves possess the commodities, for instance, assistance with infrastructure, training, and market access, are known as business development services. All parties involved in the value chain must act in an organized, systematic, and cooperative manner to provide the BDS services. It placed equal emphasis on fostering an atmosphere that is favorable to business (GFRAS, 2016).

Figure 9.2 illustrates business development services in the Agro value chain. The knowledge flows are outbound from the startups to the companies and other actors, and sometimes, in the opposite direction as well as bi-directionier, various ministries and government organizations were tasked with supporting entrepreneurship. The government decided in 2014 to create the Ministry of Skill Development and Entrepreneurship specifically to promote the growth of this industry because it believed that more government support was needed for skill development.

Below is a list of Government schemes/agencies launched to develop and encourage entrepreneurship in India.

1. Startup India Seed Fund
2. Startup India Initiative
3. ASPIRE
4. MUDRA Bank
5. Ministry of Skill Development and Entrepreneurship
6. ATAL Innovation Mission
7. eBiz Portal
8. Dairy Processing and Infrastructure Development Fund (DIDF)
9. Support for International Patent Protection in Electronics & Information

10. Technology (SIP-EIT)
11. Multiplier Grants Scheme (MGS)
12. Credit Guarantee Fund Trust for Micro and Small Enterprises (CGTMSE)
13. Software Technology Park (STP) Scheme
14. The Venture Capital Assistance Scheme (VCA)
15. Loan for Rooftop Solar Pv Power Projects
16. NewGen Innovation and Entrepreneurship Development Centre (NewGen IEDC)
17. Single Point Registration Scheme
18. Modified Special Incentive Package Scheme (M-SIPS)
19. Atmanirbhar Bharat Schemes
20. National Initiative for Developing and Harnessing Innovations (NIDHI)
21. Trade-related entrepreneurship assistance and development (TREAD)
22. Venture Capital Finance Assistance (VCA)
23. Pradhan Mantri Kaushal Vikas Yojana (PMKVY)
24. Prarambh
25. Agri Clinics and Agribusiness Centers: Better Farming by Every Farmer

Source: Author compilation.

**Figure 9.2.** Business development services in Agro value chain.

Many accelerators and incubators also play essential roles in providing mentorship, nurturing ideas, providing technical support, generating funds, and helping acquire new funds. Startup Agri-Business Incubation Program (SAIP) under RKVY-RAFTAAR in which financial assistance of a maximum of ₹25 lakhs will be granted to potential startups that have a minimum viable product (MVP) based on innovative processes/ products/ services/ business models in agriculture and allied sector. The applicant incubates would be provided funds as per their genuine requirements and the program selection committee's appraisal/evaluation of their business plans. The decision of the committee in this regard will be final. There are more than 100 agri-focused incubators in India, mostly in academic and research organizations, mainly at ICAR and agricultural universities. Schemes like Startup India, Atal Innovation Mission, NSTEDB, RKVY-RAFTAAR, and ICAR support these incubators. There are 36 agriculture-based incubation centers in India supported and promoted under the NIDHI scheme of the Department of Science & Technology (DST). Only seven TBIs are located in NARES, mainly at ICAR-IARI, New Delhi; ICAR-NDRI, Karnal; ICAR-NAARM, Hyderabad; ICAR-IIHR, Bangalore; TNAU, Coimbatore and ICRISAT, Patancheru. Similarly, there are about 45 NIDHI PRAYAS centers in India, but only one (at NAARM) in the NARES. The ICAR has set up 50 Agribusiness Incubators (ABIs) in different research institutes with an annual budget of Rs10 crores (Rs 20 lakhs each ABI). Similarly, 29 ABIs have been set up at various state agricultural universities (SAUs) under the RKVY-RAFTAAR Ministry of Agriculture & Farmers Welfare scheme with a Rs 2-3 crores grant to each R-ABI. These ABIs support over 750 agriculture-based startups and agri-entrepreneurs, including farmer-entrepreneurs and producer companies (FPOs) (NAAS,2022).

### 9.4.2. Capacity Building

Extension agencies support the sustainable development and digital transformation of agriculture by offering training and capacity-building programs to farmers, Agri-entrepreneurs, and extension workers on the use of Agri-tech tools and applications, facilitating technology adoption, encouraging collaboration, and advocacy for supportive policies. This ultimately improves productivity, profitability, and resilience throughout the industry (MANAGE).

### 9.4.3. Information Dissemination

Extension agencies provide information on the latest advancements, emerging technologies, and the best agricultural practices. Extension agents help farmers understand how Agri-tech solutions can improve productivity, efficiency, and profitability by facilitating technology demonstrations, training sessions, and field days. Extension services ensure that information about Agri-tech products, services, and market trends is reliable and relevant for farmers and agribusinesses. Information services contain details on market trends, production technologies, and market linkages. These offer a range of more information, such as the choice of market outlets and possible company associates. To guarantee the timely production of the required quantity and quality of farm food, technical information about pests, diseases, and weather patterns is needed.

### 9.4.4. On-Farm Technical Assistance

Production technologies introduced by the research system in Indian agriculture are taken to the farmer field through technology transfer and dissemination by ICAR institutes and State Agricultural Universities through the extension department. All public institutes in the Ministry of Agriculture provide agricultural support research and extension services to the farming community and tech startups in the farm ecosystem. They promote technologies such as high-value crop diversification among farmers as a risk reduction strategy and among entrepreneurs for good foreign export income. Training, workshops, and demonstrations on production information and product development are conducted. Marketing and distribution of input supplies (seeds, feeds, fertilizer, implements, machinery) provided. Awareness and knowledge of postharvest technology and storage is introduced among the stakeholders.

### 9.4.5. Off-Farm Technical Assistance

Processing, product quality, grading, and technology adoption for value-adding activities are critical aspects of value chain promotion. Agricultural extension plays a vital role in these activities by providing off-farm technical assistance to the farmers and other stakeholders by supporting services on marketing linkages and networks, assistance in market and trading centers, contract arrangements and partnership linkages, and practices provided.

Extension services fundamentally act as catalysts for the sustainable development of agriculture and the digital transformation of the agricultural industry, fostering the growth and success of agribusiness enterprises. Extension agencies are essential in stimulating innovation, collaborating with stakeholders throughout the farm ecosystem, and advocating for policies that support the growth and resilience of agri-startups. These actions contribute to the overall development of the agriculture sector and rural economy.

## 9.5. FPOs as a Key to Revolutionizing Food Value Chains

Figure 9.3 illustrates the Role of FPOs in Revolutionizing Food Value Chains. The engagement of Farmer Producer Organizations (FPOs) is crucial to promote a sustainable food value chain. Farmer Producer Organizations are one type of producer organization (the legal entity formed by primary producers, viz. farmers, the members are farmers. Producer organizations have also been established to empower communities, especially the underprivileged and other vulnerable groups, and assist in delivering services. Commercially focused farmer groups and multipurpose cooperatives concentrate on shared issues related to animal husbandry, quality control, and input supply.

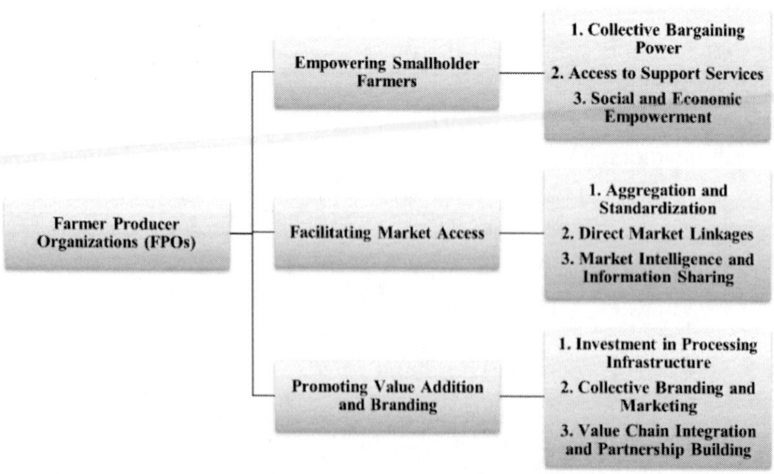


**Figure 9.3.** Role of FPOs in Revolutionizing Food Value Chains.

In addition to enhancing competitiveness, organizations can support the value chain strategy by acquiring more negotiating leverage to lower input and output transaction costs. Organizations also serve as a means of giving marginalized populations a say in public policy formulation. Organizations must be able to (i) drive change, (ii) network through established connections, and (iii) have some sort of governance to foster member trust and unity to contribute to the value chain process.

## 9.6. Challenges Faced by FPOs

- *Governance and Leadership*: Effective governance structures are crucial for FPO success. Disputes and issues related to leadership can hinder their functioning.
- *Access to Resources*: Many FPOs face challenges in accessing finance, inputs, and technology vital for modernizing agriculture.
- *Market Linkages*: Building and maintaining relationships with buyers and ensuring consistent market access can be a persistent challenge.
- *Capacity Building*: Continual training and education of members is necessary for FPOs to adapt to changing agricultural practices and market dynamics.
- *Sustainability:* FPOs often need to evolve to remain sustainable, which requires adaptability and strategic planning.

## 9.7. Initiatives by GOI in Promoting FPOs

- *Formation and Promotion of 10,000 FPOs:* The Ministry of Agriculture and Farmers Welfare launched a scheme to encourage the formation and strengthening of 10,000 FPOs across the country
- *Farmer Producer Organization Development Fund (FPDF):* Under the Ministry of Agriculture the Small Farmers' Agribusiness Consortium (SFAC) manages the FPDF to provide financial assistance to FPOs for their development and operational activities.
- *National Agriculture Market (e-NAM):* FPOs can use the e-NAM platform to trade their produce directly with buyers across different states, bypassing intermediaries and accessing better prices and markets.

## 9.8. Institutions Promoting FPOs

- *Small Farmers' Agribusiness Consortium (SFAC):* SFAC is a nodal agency under the Ministry of Agriculture and Farmers Welfare responsible for promoting agribusiness and FPOs in India
- *National Bank for Agriculture and Rural Development (NABARD):* NABARD plays a crucial role in financing and promoting FPOs through various credit and development schemes.
- *National Institute of Agricultural Marketing (NIAM):* NIAM is an autonomous institution under the Ministry of Agriculture and Farmers Welfare that focuses on research, training, and consultancy in agricultural marketing and agribusiness.

## 9.9. The Role of Extension in Mobilizing Farmer Producer Organizations to Promote Food Value Chain

Extension services are pivotal in facilitating and empowering these organizations across various value chain stages. Here's a breakdown of how extension services contribute to the mobilization of FPOs for enhancing the food value chain:

- *Production:* Extension functionaries are instrumental in promoting innovative and sustainable agricultural practices among farmers, leading to increased yields and improved incomes. For instance, initiatives like Sahaja Aaharam showcase the promotion of organic farming, and Nilokheri FPOs are involved in vegetable seed production activities to produce good quality seeds.
- *Aggregation*: Mobilizing FPOs to aggregate produce is vital to increase their bargaining power and attain better prices. Extension agents can aid FPO staff in enhancing managerial and communication skills, thus facilitating effective aggregation at the grassroots level, which is essential for optimal market positioning.
- *Processing*: Extension agents provide technical guidance to FPOs in establishing and managing primary and secondary processing units. Additionally, they provide knowledge regarding different government schemes such as the Pradhan Mantri Formalization of Micro Food Processing Enterprises Scheme (PMFME) and operation

greens to promote processing facilities at the FPO level. Arogya Millets FPO processes millets to prepare biscuits, chocolates, and ready-to-eat products

- *Distribution:* The FPOs must be trained to use digital tools such as the Kisan Rath mobile application to hire vehicles for the transportation of agricultural produce, Kisan rail services for the transportation of perishables, and Krishi Udaan for marketing agricultural produce using aviation. Even the FPOs must be encouraged to export their produce to increase their produce prices. In this regard, the extension functionaries can help FPOs follow good agricultural practices to enhance the quality standards of the produce and reduce the rejection rate in exporting. KASAM is one successful FPO that exports turmeric fingers and organic ginger slices and has received a geographical indication tag of Kandhamal Haladi
- *Marketing:* To promote better marketing opportunities for farmers FPOs must be encouraged to use digital tools such as e-NAM, AGMARKNET, and online marketing along with fostering partnerships with agri-startups and restaurants. FPOs must be involved in contract farming and retail outlet opening. The extension can also play an essential role in creating awareness of branding and developing market linkages, and communication skills for the FPO staff. Dev Bhumi Natural Products Producers Company Limited, a women-based FPO is marketing its products under their brand using digital platforms and maintaining retail outlets. Dev Bhumi Natural Products Producers Company Limited, a women-based FPO is marketing their products under their brand using digital platforms and maintaining retail outlets (Somashekhar,2014).
- *Market Research:* The FPOs should conduct market research in various market aspects, such as sales, product distribution, buyer behavior, promotion, packaging, pricing, etc., to increase the more expansive reach. Sahyadri Farms research to identify customer needs and new markets and promotes products or services that address or predict clients' objectives
- *Inputs Provision*: Extension services should facilitate the FPO's involvement in providing inputs, including seeds, organic fertilizers, finance, farm implements, knowledge, and advisory services, with some of these offered at lower prices. This approach will reduce transaction costs and build members' trust and confidence.

- *Technology Adoption*: Technology dissemination and adoption at the FPO level are higher than at the individual farmer level. Hence, this must be grabbed as an opportunity for any stakeholders in agriculture to promote high-end technologies such as the Internet of Things, drones, and innovative farming practices. The FPOs should also be enlightened regarding schemes the government has launched, such as Kisan drones to promote drones at the FPO level in which the government provides 75% grant funding. adoption at the FPO level is high compared to the individual farmer level. Hence, this must be grabbed as an opportunity for any stakeholders in agriculture to promote high-end technologies such as the Internet of Things, drones, and smart farming practices. The FPOs should also be enlightened regarding schemes that the government has launched such as Kisan drones to promote drones at the FPO level in which the government provides 75% grant funding (Fig 9.4).
- *Infrastructure Facilities:* Extension efforts should emphasize the importance of the improved marketing infrastructure and investments in logistics infrastructure. Access to cold storage systems, enhancing community farming assets, and establishing modern packaging must be done at the FPO level to reduce postharvest losses, decrease the role of intermediaries, increase the value realization for the farmers, and substantially increase their livelihood. Mahagrapes a successful FPO in Maharashtra, provides access to modern infrastructure for its members including pre-cooling and cold storage units.

**Figure 9.5.** Extension services for promoting Food Value Chain by FPOs.

Overall, extension services are a stimulant for FPOs' expansion and improvement, enabling them to reach their full potential and contribute substantially to the rural livelihood and agricultural sectors' sustainable development. Within agricultural value chains, extension initiatives create pathways for inclusive growth and economic prosperity by encouraging cooperation, innovation, and capacity building and helping them realize sustainable income and enhance livelihoods for farmers.

## Conclusion

Developing sustainable agricultural value chains is crucial for addressing various challenges within the food system. A multidimensional analysis and innovative solutions leveraging technology are vital to tackle various issues in the food system and value chain. Extension services are essential to this effort because they help farmers embrace new technologies, develop new skills, and share information. This makes Farmer Producer Organisations (FPOs) more resilient and successful. A sustainable food value chain may be achieved by combining multidimensional analysis, innovative ideas, and stakeholder cooperative efforts. This holistic approach promotes social cohesion, economic development, environmental preservation, and food security. Ultimately, with the help of extension services, the Farm to Fork initiative acts as a catalyst for creating an inclusive, resilient food system that can feed current and future generations.

## References

GFRAS. (2016). The role of extension in supporting value chains. Available at: https://www.g-fras.org/en/component/phocadownload/category/70-new-extensionist-learning-kit-nelk.html?download=648:module-10-the-role-of-extension-in-supporting-value-chains-part-1-theory (Accessed: 15 March 2024).

NAAS. (2022). Agri-startups in India: Opportunities, Challenges, and Way Forward, Policy Paper No. 108, National Academy of Agricultural Sciences, New Delhi: pp. 16.

NASSCOM. (2019). *Agritech in India - emerging trends*. https://nasscom.in/knowledge-center/publications/agritech-india-emerging-trends-2019 (Accessed: 15 March 2024).

Nuthalapati, C. S., Srinivas, K., Pandey, N., and Sharma, R. (2020). Startups with open innovation: accelerating technological change and food value chain flows in India. *Indian Journal of Agricultural Economics*, ***75***(4), 415-437.

Nuthalapati, C. S. R. (2022). Startups Digitizing Indian Food System Innovations, Survival and Investment, National Bank for Agriculture and Rural Development Research and Policy.

Sharma, J. P., and Bhatt, A. (2022). Role of agri-business entrepreneurship, innovation and value chains/networks in farmer income improvement: models, policies and challenges, *Indian Journal of Agricultural Economics*,**77**(1).

Shruti, M. Singh, and Sharma, J. P. (2018). Exploring the entrepreneurial environment for agricultural processors and developing strategies to promote value addition via Alfares' Method. *International Journal of Current Microbiology and Applied Sciences*. *7*(12), 1994-2009.

Somashekhar, I. C., Raju, J. K., and Patil, H. (2014). Agriculture supply chain management: A scenario in India. *Research Journal of social science and management*, **4**(07), 89-99.

Support for agricultural value chain development. (2012). https://www.adb.org/sites/default/files/evaluation-document/35898/files/eks-agriculturalvaluechain.pdf.

Training Programme on Supply Chain Management in Agriculture, MANAGE.

# Chapter 10

# Gamification for Sustainable Agriculture

**Pradyumna Tripathi**[1,*]
**Shaili** [1]
**and Dipika Sahoo**[2]
[1]Department of Vegetable Science, College of Agriculture,
Odisha University of Agriculture and Technology, Bhubaneshwar, Odisha, India
[2]College of Horticulture, Odisha University of Agriculture and Technology, Chiplima, Odisha, India

## Abstract

Innovation is essential in agriculture since feeding the world's population while protecting the planet's resources is a major problem. Here comes gamification, a clever strategy that has changed how we practice sustainable agriculture by going beyond the boundaries of entertainment. Using technology to encourage sustainable farming practices is known as "gamification" in the agricultural industry. It is an excellent method of involving people and adding enjoyment to the study of sustainable agriculture. To teach players about conservation, sustainable farming methods, and the value of biodiversity, farmers can consider developing interactive games or applications. It will encourage people to learn and use sustainable farming techniques by introducing competition, rewards, and progress- elements. It might include running virtual farms, choosing environmentally friendly products, and resolving issues with sustainable agriculture. It can also be utilized to enhance the fun and participatory nature of learning about sustainable agriculture in educational programs and workshops. We can encourage and enable farmers to implement sustainable practices and contribute to a greener future by utilizing gamification. Rewarding farmers for actions like effectively managing

[*] Corresponding Author's Email: sinhashaili123@gmail.com.

In: Global Horizons in Agricultural Extension Education
Editors: R. K. Doharey, Aman Verma, Satyapriya et al.
ISBN: 979-8-89530-446-4

resources or implementing new methods usually involves a mix of hardware, software, and data management systems. An intriguing use of gamification in sustainable agriculture is the idea of "virtual communities for agriculture." To cooperate, share knowledge, and exchange best practices for sustainable farming, these internet forums bring together farmers, researchers, and enthusiasts. These communities encourage member participation and motivation by implementing gamification features such as leaderboards, challenges, and badges. The adoption of sustainable methods, the provision of insightful commentary, or the accomplishment of benchmarks can all win farmer's pride. By creating a sense of community and healthy competition, this gamified approach to farming not only encourages sustainable practices but also motivates farmers to keep improving and picking up tips from each other. It is an excellent method of using gamification to move agriculture in a more positive direction. Gamification has the power to completely change how farmers think about farming and make it more gratifying and fun. With the help of gamification, sustainable agriculture may become a fun and fulfilling endeavor that promotes cooperation, learning, and group efforts to create a more environment-friendly and sustainable future.

**Keywords:** engagement, eco-friendly, scoreboard

## 10.1. Introduction

Gamification is the process of incorporating game design features and ideas into non-gaming environments to increase enjoyment, motivation, and engagement. It frequently entails adding components to non-gaming activities, such as challenges, awards, leaderboards, levels, badges, and points. Businesses, organizations, and industries can promote learning, increase participation, and improve user experience by implementing gamification approaches.

Gamification is commonly associated with the notion of games as interactive forms of entertainment that have the power to generate and augment pleasure and a feeling of accomplishment. Nevertheless, gamification may also be used for non-gaming objectives, offering a digital business environment built on guidelines and incentives to draw customers in and influence their behavior. The use of game design principles and the creation of systems to involve people, inspire actions, increase learning, and solve issues is known as gamification in the context of industrial applications. The gaming business is currently experiencing the greatest advancements in the fields of medicine, education, agriculture, tourism, entertainment, heritage, and marketing (Hamari et al., 2014) (As illustrated in Figure 10.1).

**Figure 10.1.** Shows the Overall view of Gamification.

A growing trend in agriculture is gamification, which aims to increase farmer and stakeholder involvement, productivity, and sustainability. The idea is relatively new, but it has its origins that can be traced back in agricultural extension programs that use prizes and incentives. Agricultural cooperatives' employment of incentive programs to persuade farmers to embrace new methods or technology is one historical example. These incentives could take the shape of community recognition, premium market access, or input discounts (Mollick et al., 2014). Digital technology has brought about new forms of gamification in agriculture. For example, there are apps for smartphones and websites that gamify farm management tasks like scheduling irrigation, keeping an eye out for pests, and planning (Lehtonen et al., 2017).

Gamification is becoming increasingly common and may be applied in various areas, including business, healthcare, and education. Numerous companies have successfully implemented gamified initiatives to boost employee or consumer engagement (Bogers et al., 2020).

## 10.2. Science Behind Gamification

People and companies alike frequently believe that taking action to combat climate change means sacrificing some short-term advantages in favor of longer-term ones. Psychologists have shown over decades of work that we tend to overvalue short-term gains relative to long-term ones—a process known as temporal discounting. If future benefits are viewed as being too far off, they may lose their significance or worth. To put it plainly, people frequently refuse to make short-term sacrifices or changes since they won't immediately reap the benefits (Rajkumar et al., 2016). There is often an attitude-behavior mismatch in the outcome.

People care about the environment, but they frequently don't always incorporate this care into their day-to-day actions. This similar gap explains why someone may frequently be aware of the plastic catastrophe and want to act. Still, at the same time, they may become thirsty and instinctively reach for a bottle of Coke from the vending machine without giving it any thought. It is challenging to address broad challenges like climate change, which depend on a broad shift in societal behavior, because of this disconnect between our decisions and our ideals (Bang et al., 2006). Thankfully, gamification uses little, frequent rewards to help reinforce such desirable behaviors.

## 10.3. Sustainability in Agriculture – Need of the Hour

The goal of sustainable agriculture is to meet society's current food and textile demands while preserving the capacity of present or future generations to meet their own needs. It can be predicated on an awareness of ecological services. There are numerous ways to make agriculture more sustainable. To effectively solve socio-economic and environmental issues, sustainable agriculture is crucial for several reasons.

### 10.3.1. Sustainable Agriculture is Important for the Following Principal Reasons

- *For Protecting the Environment*: Deforestation, degraded soil, water pollution, and biodiversity loss are all frequent results of conventional farming methods. Sustainable agriculture seeks to mitigate these adverse effects through the implementation of techniques that maintain ecosystems, encourage soil health, decrease the use of chemicals, and conserve natural resources (Blanchemanche et al., 2020).

- *For Protecting Biodiversity*: Ecosystem resilience and the general health of the world depend heavily on biodiversity. Sustainable agriculture, which promotes agroforestry, varied crop rotations, and the preservation of native flora and fauna, facilitates maintaining a resilient and balanced ecosystem.
- *Mitigating Climate Change*: Agriculture contributes to greenhouse gas emissions through fossil fuels, cattle raising, and deforestation. Sustainable practices—such as agroecology and organic farming—prioritize carbon sequestration, lower emissions, and increase energy efficiency to lessen this impact (Rajkumar et al., 2016).
- *Preserving Water Resources*: Conventional agriculture frequently uses a lot of chemicals and irrigation, which pollutes and depletes the water supply. Conservation tillage and precision irrigation are two examples of sustainable agricultural practices that work to lower water use, improve water quality, and increase water-use efficiency. Rainwater harvesting, recycling of wastewater and used in various agricultural processes, and the cultivation of drought-resistant crops also contribute to conserving water resources and ensuring long-term sustainability in farming.
- *Improving Soil Health*: Productive agriculture depends on healthy soils. Long-term agricultural productivity is increased by sustainable techniques that enhance soil fertility, structure, and moisture retention, such as crop rotation, cover crops, and reduced tillage.
- *Providing Food Security*: With the world's population continuing to rise, more food needs to be produced in an environmentally friendly manner. Growing productivity without sacrificing the ability of future generations to meet their food demands is the primary goal of sustainable agriculture.
- *Supporting Rural Communities*: By promoting diverse farming systems and encouraging direct consumer-producer relationships, sustainable agriculture helps to strengthen local economies. Farmers and rural communities may benefit from this through more income and better living conditions (Blanchemanche et al., 2020; Bang et al., 2006).
- *Minimizing Chemical Inputs*: Conventional farming frequently makes extensive use of artificial pesticides and fertilizers, which can be harmful to both the environment and human health. Sustainable

techniques seek to improve ecosystems and food quality by lowering chemical inputs.

- *Encouraging Resilience*: Sustainable Agricultural systems are typically more resilient to shocks and variations in the climate. A protective barrier against the effects of significant weather events and shifting conditions is provided by a variety of crops and practices that emulate natural ecosystems.
- *Long-Term Viability*: Unsustainable actions can jeopardize the ability of future generations to meet their demands by causing resource degradation and depletion. To ensure that land, water, and other resources are used in ways that can be sustained eternally, sustainable agriculture places a high priority on long-term viability.

## 10.4. Using Gamification in Sustainable Agriculture

Innovation is essential in agriculture since there are significant hurdles in feeding the world's population while protecting the planet's resources. That is until we include gamification, a clever strategy that has gone beyond fun to transform how we do sustainable farming completely. Sustainable farming practices can be encouraged through gamification, an inventive approach to agriculture (Saghaei et al., 2017). It provides farmers with information on sustainable farming practices and encourages them to implement them, all while combining the enjoyment of gaming with an educational component. Farmers are motivated to learn more about sustainability and make improvements to their farms that benefit society at large as well as themselves when the game features like points, incentives, leaderboards, and levels are incorporated. Encouraging people through gaming agricultural tasks not only helps educate but also keeps their interest longer than conventional methods of teaching, which may not be recognized for a longer period. Some points are given below to indicate the use of sustainable agriculture.

### 10.4.1. Training and Education

"To teach farmers about sustainable techniques like crop rotation, organic farming, water conservation, soil conservation, and biodiversity preservation, gamification platforms provide interactive training modules and simulations.

These platforms offer different interesting games, quizzes, and other activities that help convey complex knowledge of environmentally friendly methods in an easily understandable manner to farmers and local people" (Hanus et al., 2015).

### 10.4.2. Decision Support Systems

Farmers may maximize sustainability results by using gamified apps that offer real-time data and advice on crop management, pest control, and resource utilization.

### 10.4.3. Monitoring and Feedback

Farmers can track their adoption of sustainable practices and get immediate performance feedback through gamified interfaces, which promotes ongoing improvement (Deterding et al., 2011).

### 10.4.4. Community Engagement

By giving farmers, the opportunity to exchange stories, work together on sustainability initiatives, and take part in competitions with the shared objective of protecting the environment, gamification helps to build a feeling of community among farmers.

### 10.4.5. Mechanisms for Incentive

Leaderboards, badges, and rewards encourage farmers to implement and uphold sustainable practices, promoting long-term environmental stewardship and behavior change.

### 10.4.6. Data Collection and Analysis

Farming practices, environmental impact, and productivity are all subject to important data collected by gamification platforms. This data may be examined to spot trends, judge the success of initiatives, and guide policy decisions. It also helps in the easy analyzing of data that allows improving yields, minimizes wastages of resources, and decrease impact of environment (Oinas –kukkonedn et al., 2009).

### 10.4.7. Behaviour Change

Gamification might encourage farmers to embrace sustainable practices by creating challenges and offering prizes to meet predetermined objectives. For example, farmers could receive badges or points for adopting agroforestry, decreasing their use of chemicals, or managing their water resources well.

### 10.4.8. Precision Farming or Site-Specific Farming

Gamification can be included in GPS-guided devices and sensors, among other precision agriculture technology. To encourage more effective use of resources, farmers may be rewarded with points for precisely planting, fertilizing, and irrigating their crops, which ultimately reduce the cost of production (Seaborn et al., 2015).

### 10.4.9. Crop Management

Farmers can test out various crop rotation plans and cover cropping tactics and soil management strategies with the aid of virtual farming simulations and management games (Thompson et al., 2016). They may comprehend the long-term advantages of sustainable activities without having to take any actual risks because of this.

### 10.4.10. Supply Chain Transparency

Customers can discover the provenance and manufacturing process of the food they eat by using gamified traceability solutions. This might stimulate demand for products made responsibly, which would then motivate farmers to use environmentally friendly methods.

### 10.4.11. Reducing Carbon Footprint

Using gamified carbon footprint calculators and challenges can motivate farmers to adopt greenhouse gas emission-reducing techniques. This can entail enhancing nutrition management, implementing sustainable energy sources, and streamlining transportation.

### 10.4.12. Reducing Waste

Farmers can be encouraged to reduce waste and advance circular economy principles by using gamification. For instance, challenges might be created to promote composting organic matter recycling or creative uses of agricultural wastes (Rapp et al., 2019).

### 10.4.13. Policy Advocacy

Farmers and advocates can interact with legislators and spread the word about the value of sustainable agriculture by using gamified platforms. Changes in policy that encourage and support eco-friendly behavior may result from this.

### 10.4.14. Soil Health Monitoring

Gamified soil testing and monitoring tools can encourage farmers to regularly assess soil health parameters such as nutrient levels, pH balance, and organic matter content. Interactive feedback and recommendations can guide farmers toward sustainable soil management practices.

Gamification has the potential to be a potent instrument for advancing sustainable agriculture by motivating and enabling farmers to adopt eco-friendly practices, preserve natural resources, and guarantee the long-term sustainability of farming systems.

## 10.5. Tools and Features Incorporated in Gamification for Sustainable Agriculture

A variety of tools and features are used in gamification for sustainable agriculture to inform, engage, and inspire users to adopt ecologically friendly farming methods. Sustainable agriculture can be made more exciting by gamification by adding the following features:

- *Educational Modules*: Crop rotation, organic pest control, soil conservation, and water conservation are just a few of the sustainable farming practices that users may learn about using gamified platforms (Hanus et al., 2015).
- *Points and Incentives*: Farmers can exchange their points for discounts or incentives by using them to encourage sustainable actions like soil regeneration or water conservation.
- *In-game Currency*: Farmers earn virtual currency or points by achieving sustainability goals, which can be used to unlock new features, buy virtual goods, or compete with other players (Rapp et al., 2019).
- *Leaderboards*: Showcasing the best performers promotes healthy rivalry and inspires farmers to use sustainable practices.
- *Challenges & Quests*: Establishing specific objectives and challenges, like cutting back on pesticide use or diversifying crop varieties, implementing cover cropping, or adopting an integrated nutrient management system, helps farmers stay motivated and focused on sustainability.
- *Progress racking:* Keeping track of one's progress towards sustainable objectives enables farmers to make necessary adjustments to their methods and stay on course.
- *Progressive Challenges*: Users may be given tasks pertaining to sustainable agriculture that are gradually more difficult on gamified

platforms. Users get incentives and a sense of achievement when they do activities reach milestones and unlock new content.

- *Virtual Farms*: Farmers can experiment with sustainable practices without endangering their genuine crops by building virtual versions of their actual farms. Firstly, they will create a virtual representation of their real-world farm, where they can easily experiment with their different sustainable practice ideas (Oinas-Kukkonen et al., 2009). Not only this, but they can also quickly check the potential impact on their farm's productivity and sustainability.

*Example:* - "FARMING SIMULATOR" and "HARVEST: MASSIVE STIMULATOR"

- *Instructional Content*: Farmers' knowledge and abilities are improved by offering tutorials and instructional materials on sustainable methods.
- *Social Sharing*: Encouraging farmers to use social media to share their successes and setbacks builds a community of support and raises awareness of sustainable agriculture.
- *Storytelling and Narratives*: Gamification can incorporate storytelling elements to create immersive experiences that educate and inspire farmers about the importance of sustainable agriculture and their role in preserving the environment for future generations.
- *Real-time Monitoring and Feedback*: Integration with IoT devices and sensor networks enables real-time monitoring of environmental conditions and farming activities. This allows farmers to receive immediate feedback on their progress and adjust as needed. By incorporating real-time data analysis technologies, farmers can keep an eye on and measure the environmental effects of their operations, including greenhouse gas emissions, water use, and soil health. To maximize their sustainability efforts, consumers can make well-informed decisions with this input. Forums, chat rooms, and social media integration are examples of elements that are frequently seen in gamified platforms that help to foste
- *Customization and Personalization*: To maintain relevance and efficacy, gamification features can be tailored to suit varying farm sizes, crops, and climates. Farmers can customize their gaming experience by selecting their preferred sustainable practices, setting

their own goals, and receiving personalized feedback and recommendations based on their performance.

- *Social Collaboration*: Farmers can collaborate with peers, experts, and agricultural extension services through social features like chat, forums, and multiplayer challenges, fostering knowledge exchange and collective problem-solving (Mentioned in figure 10.2)
- *Prizes & Incentives:* Encouraging users to actively engage in sustainable activities by providing them with prizes and incentives, like in-game goodies, virtual currency, or badges, encourages them to start acting sustainably.

Sources: ecologi.com.

**Figure 10.2.** Shows social collaboration.

## 10.5.1. FARM BOT

In gamification, a farm bot could be a feature or tool designed to automate specific tasks or actions within a game that simulates farming or resource management. Farm bots can be used to gamify virtual settings and games to produce captivating and immersive farming experiences. Here's how to do it:

For example, in a farming simulation game, a farm bot might automatically plant, water, and harvest crops for the player, allowing them to progress in the game without having to perform those actions manually. It adds an element of efficiency and convenience, making the gameplay experience more enjoyable for some players. However, game developers need to balance the use of such bots to ensure that they don't undermine the core gameplay or challenge too much. To help farmers improve their agricultural methods, the system also offers real-time data and comments. Framing games with farms also improves farmers' knowledge related to agriculture, such as teaching players about crop rotation, soil health, and most crucial sustainable agriculture (Conaway et al., 2014). Farm Bot encourages farmers to actively participate in sustainable and productive agricultural practices by introducing gamification features like achievements and progress tracking, players can earn rewards by competing through games based on having the most efficient or productive farm.

"A genuine instance of a farm bot in gamification may be found in the well-known mobile game "Hay Day." In Hay Day, users cultivate and expand their farms by rearing livestock, planting crops, and making products (Saghaei et al., 2017). To assist players in running their farms more profitably, the game provides several improvements and tools, including devices that can automate chores like feeding the animals and harvesting crops. These "farm bots" improve the overall gaming experience by allowing players to advance more quickly and accomplish their farming objectives with less human labor (Figure 10.3)."

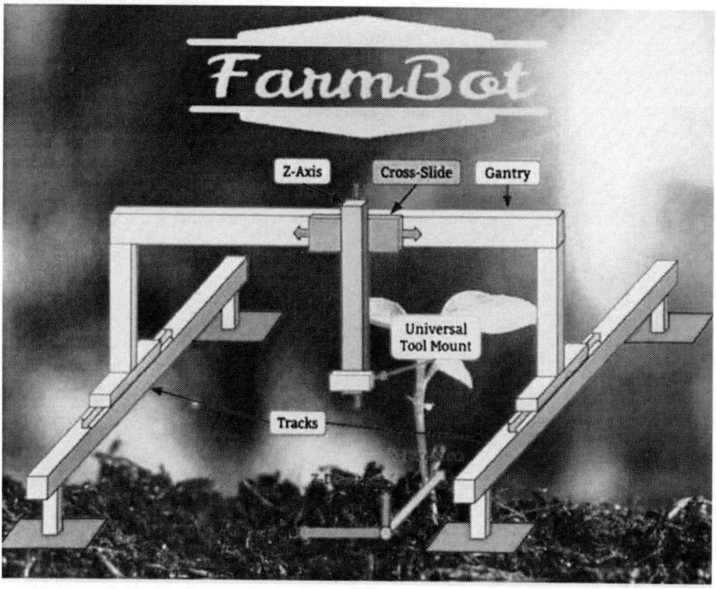


Sources: blog.techdesign.com.

**Figure 10.3.** Shows farmbot App.

## 10.6. Examples of Gamification in Sustainable Agriculture in the Real World

### 10.6.1. Plantix

Plantix is a mobile app that uses image recognition technology to identify crop diseases and pests. It gamifies the learning process by allowing farmers to earn points for correctly identifying issues and providing them with recommendations for sustainable solutions. With Plantix, you can quickly and reliably identify pests and diseases on crops using your Android phone as a mobile crop doctor. Plantix offers a comprehensive solution for managing and producing crops (Conaway et al., 2014). The Plantix application includes coverage for and recognises—only by snapping a picture of—a diseased crop. It can be found in and has received more downloads than This results in for identifying damage, managing pests and diseases, and enhancing crop productivity for farmers across the globe. It will Find agricultural pests and illnesses and give suggested solutions. Plantix is a mobile app that uses image recognition technology to identify crop diseases and pests. It gamifies the learning process by allowing farmers to earn points for correctly identifying issues and provides them with recommendations for sustainable solutions. With Plantix, you can quickly and reliably identify pests and diseases on crops using your Android phone as a mobile crop doctor. Plantix offers a comprehensive solution for managing and producing crops (Conaway et al., 2014).

The Plantix application includes coverage for and recognizes a wide range of crops—only by snapping a picture of—a diseased crop. It can be found in over 150 countries and has received more than 10 million downloads. This results in quicker, more accurate identification of crop damage, effective pest and disease management, and enhanced crop productivity for farmers across the globe.

It will find agricultural pests and illnesses and give suggested solutions based on real-time data, improving decision-making and promoting sustainable farming practices.

### 10.6.2. Farm Bucks

Farm Bucks is a rewards-based program that incentivizes farmers in India to adopt sustainable practices. Farmers earn points for using organic fertilizers, implementing water-saving techniques, and reducing pesticide use. These points can be redeemed for discounts on agricultural inputs (Deterding et al., 2011).

It has different features that lead to sustainable agriculture

- *Management of Farm*: Players can virtually manage resources related to the farm like raising animals, cultivating crops, and managing resources. This same feature is also in the game FarmVille
- *Marketplace*: Here, players can virtually sell and buy goods related to farming. Prices are not constant here; it fluctuates according to the supply and demand of goods. Similar mechanics can be found in a game like "Stardew Valley."
- *Social Interactions*: It allows farmers to visit and interact with each other, share their ideas and knowledge, collaborate for work and sometimes exchange gifts. These types of features prevail in game Hay Day.

### 10.6.3. TruTrade

TruTrade, a platform in Africa, uses gamification to engage smallholder farmers in sustainable agricultural practices. The platform allows farmers to earn points for producing high-quality crops, which can then be exchanged for access to credit, training, and other resources

A "TruTrade" gamification platform for agriculture may have the following features:

- *Market Simulation:* This game allows users to purchase, sell, and trade different crops and commodities according to market conditions. It simulates real-world agricultural markets. Resource management: To maximize crop yield and profitability, manage resources like land, water, seeds, and fertilizer.
- *Supply Chain Management*: Gain knowledge of the complete agricultural supply chain, from seeding to crop harvesting to market

transportation. Risk management: Players must navigate obstacles like bad weather, pests, and market swings while making calculated judgments to reduce risks and guarantee success (Mollick et al., 2014).

- *Instruction and Guidance*: Offer players instructional resources and guides to assist them in learning about trading tactics, sustainable farming methods, and agricultural economics.
- *Playing Games in Competition*: Play against AI opponents or other players to become the best farmer or agricultural merchant in the game's competitive mode. Gain access to data analytics and insights to monitor performance, spot trends, and make wise judgements.
- *Community Interaction*: Promote community interaction so that players may work together, exchange advice, and talk about winning gaming techniques.

### 10.6.4. Plant Village Nuru

Plant Village Nuru, an AI-powered app created by Penn State University, assists farmers in identifying and treating crop illnesses. By giving badges and prizes to users who actively offer knowledge and support other users on the network, it combines gamification features.

"Plant Village Nuru" has a captivating sound! It appears to be a gamified method of practicing sustainable agriculture. The idea of adding game components to agriculture has the potential to increase public awareness of sustainable farming methods and to make them more enticing and approachable (Seaborn et al., 2015). This could involve using a virtual environment to manage crops, water use, soil health, and insect control, among other things. Learning about the significance of sustainable agriculture and its effects on the environment and food security may be entertaining and engaging.

Some of its features are given below:

- *Crop Management*: A variety of crops, each with unique growing requirements and challenges, are available for players to select and produce. Soil health involves keeping an eye on the quality of the soil and putting policies like cover crops and crop rotation into place to make the soil healthier.

- *Water Management:* Players need to effectively control the amount of water they use, considering things like irrigation strategies and water-saving measures.
- *Pest and Disease Control*: Using integrated pest management techniques, such as organic pesticides and natural predators, to identify and eradicate pests and diseases. Agroforestry, composting, and minimum tillage are a few examples of sustainable agricultural techniques that players are encouraged to embrace.
- *Weather Events:* Managing erratic weather conditions such as floods, storms, and droughts that can affect crop production and call for adaptable techniques. Selling crops in a dynamic market where supply and demand the state of the weather (Bang et al., 2006).

### 10.6.5. Rice Advice

The International Rice Research Institute has created a smartphone application called Rice Advice (IRRI). It offers customized advice on how to increase yields and sustainability to rice producers in Asia and Africa. To keep farmers interested and make learning about best practices pleasurable, the app incorporates interactive components. This free Android app offers guidance on rice management techniques tailored to individual farms. This app can be used by farmers, extension agents, the commercial sector, and development organizations who own Android smartphones or tablets. Farmers must respond to questions about their rice-growing conditions, crop management, current production level, etc. to create the guidelines (Barab et al., 2009). This interview takes five to ten minutes Based on nutrient requirements and fertilizer prices, Rice Advice can determine which fertilizers are best to buy, in what quantities, and when to use them. Farmers can choose their own desired yield level using the software, considering their financial constraints. Except for software updates, Rice Advice does not require an Internet connection to generate the guidelines. But it has severe issues. The issue is that many farmers in Sub-Saharan Africa lack access to smartphones to access and utilize Rice Advice. Finding appropriate commercial structures for offering Rice Advice could solve this. A business strategy that combines the provision of Rice Advice services with other agricultural services (such as seed, machinery, etc.) is being tested (Figure 10.4). You can download the Rice Advice app for free on any Android device (tablet or smartphone) from the Google Play Store. There are versions accessible in French and English, and more languages are on the horizon (Deternding et al., 2011).

Source: africarice.org.

**Figure 10.4.** Show Rice Advice App symbol.

### 10.4.6. Ava's Quest for Good Crops by Corteva Agriscience

For Quality Vegetables by Corteva Agriscience: An instructional game called Ava's Quest teaches young pupils about food production, sustainability, and agriculture. Ava Quest is an app that makes learning about agriscience and agriculture more fun. It offers a range of interactive modules and quests. Several typical characteristics are as follows:

- Users can learn about a variety of agricultural topics, including crop cultivation, insect control, soil health, and more, by completing quests—engaging assignments.
- *Mini Games:* Educative and entertaining games that highlight important ideas in agriculture and agriscience.
- *Materials:* Access to enlightening publications, films, and additional materials to enhance knowledge of agricultural subjects.
- *Community:* Possibilities for user interaction, experience sharing, and teamwork on farming initiatives.

- *Progress Tracking*. Within the app, there are tools to track your accomplishments and progress—updates and difficulties, challenges and frequently updated content to keep visitors interested and inspired to keep studying. For users of all ages, these features come together to create an immersive learning environment that makes agriculture and agriscience approachable and entertaining (Mollick et al., 2014).

### 10.4.7. Farmville Green Acres Initiative

To start the Green Acres Initiative, popular social farming game Farmville teamed up with a number of environmental organisations. Engaging in virtual ecological tasks, such as planting trees, conserving water, and using sustainable farming methods, was recommended for players (Gerling et al., 2018). Farmville made a significant difference outside of the virtual world by contributing money to actual sustainability projects for each challenge that was successfully finished.

Zynga created the well-known farming simulation social network game FarmVille. Among its attributes were:

- *Farming:* On their virtual farms, players could grow crops, plants, and livestock.
- *Decorations:* Adding different structures, landscape elements, and decorations to personalize the farm.
- *Social Interaction*: Players might go to the farms of their neighbours, provide a hand, and trade gifts
- *Quests and Achievements*: Reaching new levels and earning rewards requires completing tasks and challenges (Lehtonen et al., 2017).
- *Expansion:* Increase the size of the farm as the player advances by opening more rooms and features.
- *Cooperative Play*: Collaborative chores and activities are made possible by cooperative farming with pals.
- *Marketplace:* Purchasing and selling products within the game. Events: Special incentives for limited-time promotions and seasonal events.
- *Customization:* Adding unique touches to the farm's design, layout, and avatar.

### 10.4.8. Agri Craft

University of Washington academics created the mobile game Agri Craft with the goal of educating users about sustainable farming methods. In a virtual farm management game, players choose crops, set irrigation schedules, and deal with pests. The game includes instructional material about sustainable agricultural methods to help players comprehend the advantages these methods have for the environment and the economy. Agri craft is a mod that improves agriculture in the video game Minecraft (Perrotta et al., 2016). Crop breeding, which allows you to crossbreed plants to produce new ones with distinctive properties, irrigation systems for automated watering, and growth accelerators for accelerated plant growth are a few key aspects. It also provides compatibility with other mods for additional features and greenhouse constructions that may be customized. Here are a few more crucial aspects of agri craft to consider:

- *System of Mutations*: Agricraft presents a technique of plant mutation that enables the production of specialized and diverse crops by allowing plants to mutate into new types under certain conditions.
- *Genetic Engineering*: Using gene modification tools and mechanics, players may change the genetics of their crops, changing things like growth rates, yield, and tolerance to environmental influences.
- *Seed Storage*: With the help of the mod's seed storage system, players may effectively arrange and store their seeds, making large-scale farming operations easier.
- *Compatibility:* Agricraft is made to work with other mods. This allows it to be integrated with other agricultural and resource management systems and allows different modded plants to breed with one another (Hamari et al., 2014). Resource Generation: By using sophisticated agricultural methods and automation, players can use agricraft to produce renewable resources like food, materials, and even energy. Research and Achievement: By allowing users to experiment and do research, the mod encourages them to try out various breeding combinations and improve their agricultural setups.
- *Community and Support*: Via forums, wikis, and other online tools, a committed group of players and developers actively contribute to the growth of Agricraft by helping, updates, and new material.

### 10.4.9. Farming Simulator

A sustainable agriculture-focused expansion pack for the well-known simulation game Farming Simulator was released. Organic farming, sustainable energy sources, and environmentally friendly machinery are just a few of the new gameplay options available to players (Rajkumar et al., 2016). Players may reduce the environmental effect of their virtual farm while increasing its yield by putting these sustainable ideas into practice (Figure 10.5).

A farming simulator app has the following features:

- *Crop Management*: Users can plant, grow, and harvest a variety of cropsthat. Handle the health and productivity of animals such as pigs, chickens, and cows when raising them as livestock.
- *Equipment Management*: Invest in and keep up agricultural machineries such as harvesters, plows, and tractors.
- *Market Interaction*: Bargain with buyers and sell agricultural and animal products at market rates.
- *Weather System*: Be aware of shifting weather patterns that have an impact on crop growth and harvest dates. Farm Expansion: Make an investment to buy more land, create infrastructure, or extend your farm.
- *Missions and Challenges*: Fulfill objectives and challenges to get rewards, such as meeting output goals or surviving natural calamities.
- *Community Engagement*: Engage in trade items, socialize with other gamers, or take part in farming contests.
- *Realistic Simulation*: Precise farming mechanisms, such as fertilization, pest control, and soil management.
- *Seasonal Events:* Exclusive challenges and awards that are linked to actual seasons and holidays in the real world.

### 10.4.10. Life Company's Reforestation App

A smartphone app created by Land Life Company makes the reforestation process more entertaining. Features of the Life Company reforestation app are as follows:

- *Tree planting tracker*: Users can keep tabs on the growth of the trees they have planted. Provide users with the option to sponsor the planting of trees or make donations to forestry efforts.
- *Education hub:* Offer materials outlining the value of environmental preservation and replanting.
- *Community involvement*: Encourage user conversations, advice exchanges, and success stories. Impact calculator: Display to users the effects of their tree-planting activities on the environment.
- *Suggestions for planting according to location*: Provide native tree species that are appropriate for the users' locations.
- *Progress badges*: Encourage users to keep participating by rewarding them when they meet planted milestones (Lehtonen et al., 2017).

Users can follow the development of their planted trees and take part in online tree-planting competitions. Users are encouraged to feel engaged and accomplished by the app, which gives them real-time updates on the growth of the trees and their impact on the environment. Additionally, by purchasing virtual trees within the app, users may support actual forestry initiatives.

Source: https://www.aptoide.com/app/com.giantssoftware.fs14/farming-simulator-14.

**Figure 10.5.** Shows farming simulator game.

## 10.5. Benefits of Using Gamification in Agriculture

There are many advantages to gamification in agriculture, including improved farmer behavior, knowledge retention, and engagement. Agricultural processes can be made more sustainable, encourage learning, and enhance decision-making abilities by using game principles. According to research by (J. Hamari, J. Koivisto, & H. Sarsa 2014), gamification can improve engagement and motivation, which will enhance the uptake of agricultural innovations. Furthermore, gamification can assist in data collection for research, as demonstrated by initiatives such as the Farmville game on Facebook, which collected information on actual farming methods (Thompson et al., 2016). Furthermore, as evidenced by websites like Farmingville, where users can share ideas and experiences, it can encourage cooperation and knowledge-sharing among farmers (Saghaei et al., 2017). Gamification has the potential to transform agricultural education and practice and education, leading to more productive and sustainable farming techniques practice and education, leading to more productive and sustainable farming techniques.

## 10.6. Prospects of Gamification for Sustainable Agriculture

Gamification holds promising potential for sustainable agriculture by incentivizing eco-friendly practices among farmers, fostering knowledge sharing, and promoting community engagement. It can encourage the adoption of precision farming techniques, conservation practices, and sustainable resource management. Additionally, gamification can help educate consumers about the environmental impact of food choices, fostering a more sustainable food system overall. As technology advances, integrating gamification with data analytics and artificial intelligence could further enhance its effectiveness in promoting sustainable agriculture practices (Barab et al., 2009).

Here are some more details about the potential applications of gamification in the future:

### 10.6.1. Increased Acceptance

Gamification is expected to witness a rise in acceptance as digital platforms proliferate in the agricultural industry. Gamified apps and platforms have the potential to increase learning and accessibility of sustainable practices for farmers and agricultural organizations.

### 10.6.2. Personalisation and Customisation

Upcoming gamification initiatives might concentrate on adjusting experiences to suit the requirements and tastes of certain users. Personalized challenges, prizes, and feedback systems based on variables like geography, farm size, and particular sustainability objectives might be included.

### 10.6.3. Integration with Developing Technologies

Gamification could be combined with developing technologies like augmented reality (AR) and virtual reality (VR) to create immersive and engaging experiences. Farmers may, for instance, utilize augmented reality apps to see how various farming techniques.

### 10.6.4. Cross-Sector Collaboration

By bringing in a variety of viewpoints, resources, and areas of expertise, working with organizations, government agencies, businesses, and academic institutions can enhance gamification projects and promote innovation and scalability (Conway et al., 2014)

### 10.6.5. Impact Scaling

By extending the reach of gamification programs to a broader range of farmers and stakeholders, collaborations with financial institutions, market intermediaries, and providers of agricultural inputs can enhance their impact.

### 10.6.6. Incentive Design

Behavioral economics principles can be used to design incentive structures that effectively inspire farmers to participate in sustained behavior change. Examples of these structures include gamified rewards, social recognition, and status symbols.

### 10.6.7. Long-Term Impact Assessment

To optimize efficacy and refine tactics, it is imperative to implement strong monitoring and evaluation procedures to determine the long-term impact of gamification on sustainable agriculture outcomes, including environmental, social, and economic dimensions.

### 10.6.8. Policy Integration

By incorporating gamification techniques into more comprehensive development agendas and agricultural policies, it is possible to guarantee that these will be in line with global and national sustainability goals, which will promote systemic change and mainstream sustainable practices in the agricultural industry (Deterding et al., 2011). Gamification has the potential to significantly impact sustainable agriculture and support international efforts to combat climate change, biodiversity loss, and food hunger by adopting these cutting-edge tactics.

## Conclusion

In this chapter, we have discussed gamification in sustainable agriculture, as gamification depicts the game for entertainment. Here, we are practicing sustainable agriculture by going beyond the boundaries of entertainment. It is an emerging concept that will improve sustainability and maintain ecological balance. It is an excellent method of involving people and adding enjoyment to the study of sustainable agriculture. To teach players about conservation, sustainable farming methods, and the value of biodiversity, farmers can consider developing interactive games or applications. It will encourage people to learn and use sustainable farming techniques by introducing competition, rewards, and progress-tracking elements. It might include running virtual farms, choosing environmentally friendly products, and resolving issues with sustainable agriculture. It can also utilize the fun and participatory nature of learning about sustainable agriculture in educational programs and workshops. We can encourage and enable farmers to implement sustainable practices and contribute to a greener future by utilizing gamification (Hamari et al., 2014).

Gamification in sustainable agriculture involves applying game design principles and mechanics to engage farmers and stakeholders in environmentally friendly practices. It fosters knowledge sharing, encourages competition in a positive way, and motivates participants to contribute to sustainable agriculture efforts. Foer example: farmers might earn points and badges for using eco-friendly farming methods to achieve specific sustainability goals, creating a sense of achievement and camaraderie among participants.

## References

Bang, M., Torstensson, C., & Katzeff, C. (2006). The powerhouse: A persuasive computer game designed to raise awareness of domestic energy consumption. *International conference on persuasive technology* (pp. 123–132). Berlin: Springer.

Barab, S., Thomas, M., Dodge, T., Carteaux, R., & Tuzun, H. (2009). Making Learning Fun: Quest Atlantis, A Game Without Guns. *Educational Technology Research and Development,* 55(6), 537–579. doi:10.1007/s11423-007-9072-8.

Blanchemanche, S., Garedakis, N., & Mathé, S. (2020). Gamification in Agriculture: A Review of Current Tools and Opportunities. *Sustainability,* 12(3), 1123.

Bogers, M., & van den Bulte, C. (2020). The Business of Platforms: Strategy in the Age of Digital Competition, Innovation, and Power. *Harvard Business Review Press.*

Conaway, R., & Garay, M. (2014). *Gamification and service marketing*. Springer Plus, 3(1), 653.

Deterding, S., Dixon, D., Khaled, R., & Nacke, L. (2011). From Game Design Elements to Gamefulness: Defining "Gamification". In *Proceedings of the 15th International Academic MindTrek Conference: Envisioning Future Media Environments* (pp. 9–15). ACM. doi:10.1145/2181037.2181040.

Gerling, K. M., Klauser, M., & Oehlberg, L. (2018). Design strategies for gamified physical activity applications for children with cerebral palsy. In *Proceedings of the 2018 CHI Conference on Human Factors in Computing Systems* (pp. 1-14).

Hamari, J., Koivisto, J., & Sarsa, H. (2014). Does gamification work? A literature review of empirical studies on gamification. In *Proceedings of the 47th Hawaii International Conference* on System Sciences (pp. 3025-3034).

Hanus, M. D., & Fox, J. (2015). Assessing the Effects of Gamification in the Classroom: A Longitudinal Study on Intrinsic Motivation, Social Comparison, Satisfaction, Effort, and Academic Performance. *Computers & Education,* 80, 152–161. doi: 10.1016/j.compedu.2014.08.019.

Lehtonen, M., Kutvonen, L., & Smed, J. (2017). A review of the literature on gamification in agriculture. *Computers and Electronics in Agriculture,* 142, 146-158.

Mollick, E., & Rothbard, N. (2014). Mandatory Fun: Gamification and the Impact of Games at Work. *Games and Culture,* 9(3), 202–227. doi:10.1177/1555412013519606.

Oinas-Kukkonen, H., & Harjumaa, M. (2009). Persuasive Systems Design: Key Issues, Process Model, and System Features. *Communications of the Association for Information Systems,* 24, 485–500. doi:10.17705/1CAIS.02428.

Perrotta, C., & Cimino, M. G. (2016). Gamification and education: A literature review. In *International Conference on Immersive Learning* (pp. 173-189). *Springer,* Cham.

Rajkumar, S., & Ramaswamy, K. V. (2016). Gamification of agricultural extension: A study on the sustainable adoption of integrated pest management practices. *Computers and Electronics in Agriculture,* 127, 321-333.

Rapp, A., Hopfgartner, F., Hamari, J., Linehan, C., Cena, F., & Streitz, N. (2019). Introduction to the Special Issue on Gamification and Serious Games for Learning. *International Journal of Human-Computer Studies,* 127, 1–6. doi: 10.1016/j.ijhcs. 2019.02.003.

Saghaei, A., Chang, S., Tang, S., & Sulaiman, M. (2017). Design and Implementation of Serious Games for Agriculture: A Systematic Literature Review. In *2017 6th ICT International Student Project Conference* (ICT-ISPC).

Seaborn, K., & Fels, D. I. (2015). Gamification in Theory and Action: A Survey. *International Journal of Human-Computer Studies,* 74, 14–31. doi: 10.1016/j.ijhcs. 2014.09.006.

Thompson, A., Coravos, A., & Marsden, G. (2016). *Farmville: Past, Present, and Potential Futures of Playful Data.* In *Proceedings of the 2016 CHI Conference Extended Abstracts on Human Factors in Computing Systems.*

University of Washington Center for Public Health Nutrition. (n.d.). *FarmCraft.*

Zichermann G., & Cunningham, C. (2011). Gamification by Design: Implementing Game Mechanics in Web and Mobile Apps. O'Reilly *Media*, Inc.

Chapter 11

# Artificial Intelligence and Big Data in Extension: Leveraging Technology for Informed Decision-Making

**Ramesh Chand Bunkar[1,*]**
**Gopal Sankhala[1]**
**Laksheeta Chauhan[2]**
**Aman Verma[3]**
**M. Vamshi[1]**
**Sunil Kumar Meena[5]**
**Devendra Singh[2]**
**and Rishabh Singh Gaur[1]**
[1]Division of Dairy Extension, Indian Council of Agricultural Research- National Dairy Research Institute, Haryana, India
[2]Rajasthan College of Agriculture, Maharana Pratap University of Agriculture and Technology, Rajasthan, India
[4]Department of Extension Education, Acharya Narendra Deva University of Agriculture and Technology, Ayodhya, Uttar Pradesh, India
[5]Sri Karan Narendra Agriculture University, Jobner, Jaipur, Rajasthan, India

## Abstract

Extension services are vital in providing farmers and agricultural stakeholders with information, resources, and support to improve their productivity and livelihoods. Informed decision-making is essential for

---

* Corresponding Author's Email: rameshndri.icar@gmail.com

In: Global Horizons in Agricultural Extension Education
Editors: R. K. Doharey, Aman Verma, Satyapriya et al.
ISBN: 979-8-89530-446-4

optimizing agricultural practices and mitigating risks in an ever-changing environment. Leveraging artificial intelligence (AI) and big data presents unprecedented opportunities to enhance the effectiveness of extension services by providing timely, data-driven insights and recommendations. This article explores the intersection of AI and big data in extension services and its potential impact on informed agricultural decision-making. It discusses critical applications of AI and big data in precision agriculture, farmer advisory systems, pest and disease management, and climate resilience. It also examines the benefits, challenges, and best practices of integrating AI and big data into extension services. Through case studies and examples from around the world, it highlights successful implementations and lessons learned. To optimize the potential of technology in empowering agricultural communities and fostering sustainable development, continuing adaptation and cooperation are also important. Finally, the article discusses recommendations for reveraging technologies to enhance decision-making in extension services.

**Keywords:** Artificial Intelligence, big data, decision-making, extension services, leveraging, technology

## 11.1. Introduction

Artificial intelligence (AI) and big data have emerged as transformative tools in extension services, revolutionizing information gathered, analyzed, and utilized for informed decision-making. Through AI algorithms and machine learning techniques, extension agents can sift through vast amounts of data collected from various sources, including satellite imagery, weather sensors, and historical records, to identify patterns, trends, and insights that were previously inaccessible. This enables more accurate predictions of agricultural outcomes, such as crop yields, pest infestations, and soil health, empowering farmers and policymakers to make timely and evidence-based decisions (Smith, 2023).

Furthermore, AI-driven extension services can personalize recommendations and interventions to individual farmers, considering their specific needs, constraints, and preferences. By leveraging AI-powered chatbots or mobile applications, farmers can access real-time guidance and advice tailored to their crop type, location, and current conditions, facilitating proactive problem-solving and optimizing resource allocation. Moreover, big data analytics can enhance collaboration and knowledge-sharing within agricultural communities, enabling stakeholders to exchange best practices, lessons learned, and innovative solutions for sustainable agriculture. Ultimately, integrating AI and big data in extension services can increase productivity, resilience, and inclusivity in agricultural systems, contributing to food security and rural development worldwide (Smith, 2023).

## 11.2. Artificial Intelligence

Artificial intelligence (AI) is an interdisciplinary branch of study that models human actions and cognition, including learning and problem-solving, to replicate human intellect in robots. Extension specialists and research scientists are utilizing Al technology to solve problems related to agricultural production and productivity. AI can help farmers increase yields by guiding them in choosing suitable crop varieties, implementing improved soil and nutrient management strategies, managing pests and diseases, estimating agricultural productivity, and anticipating commodity prices. Al uses cutting-edge methods such as deep learning, machine learning, IoT, robots, image processing, artificial neural networks, and wireless sensor networks (WSN) to solve agricultural challenges. These Al technologies may now assist farmers in monitoring various variables collected from their fields in real-time, such as weather, temperature, water usage, or soil conditions, to enhance decision-making. Al is used to innovative farming methods that minimize losses for farmers while providing superior crops (Doe, 2024). Al is a branch of computer science that simulates human intelligence by learning from and interpolating data using various methods, such as machine learning and deep learning. To provide predictions, these networks dynamically link input and output variables. These predictions can be used to generate a range of solutions for issues that are both straightforward and difficult. Al-powered electronics are already widely used in our daily lives. AI is pervasive today, from self-driving cars to smartphone face recognition software. es. Thanks to AI systems and ML expertise, agriculture will never undergo a digital transformation despite other industries witnessing significant productivity increases, thanks to AI systems and ML expertise. Nevertheless, AI is bringing one of the oldest industries up to date. AI finds extensive application in agriculture. Precision farming is made possible by artificial intelligence. AI may help farmers with irrigation, crop rotation, harvesting, crop selection, planting, and pest control by using data from ML (Doe, 2024).

## 11.3. Types of Artificial Intelligence

There has been a consistent increase in food insecurity globally since 2019; by 2030, 670 million people, or 8% of the world's population, are predicted to be hungry despite implementing the 2030 Agenda in 2015 and several programs to end hunger. Food insecurity results from a lack of resilience in most food systems and is s exacerbated by population growth. Artificial intelligence applications enhance availability, utilization, stability, and accessibility (Johnson, 2023).

The principal streams of AI include Machine Learning (ML), Computer Vision, Natural Language Processing, Artificial Neural Networks, Robotics and Expert Systems.

### 11.3.1. Machine Learning (ML)

Machine Learning (ML), a subset of AI, has shown great potential in enhancing various aspects of Agriculture 4.0. It can be defined as a computer program or system that can learn specific tasks without being explicitly programmed. It involves using a computer to make decisions based on multiple data inputs. Labeled data is often used for supervised learning tasks (where the model learns from labeled examples), and unlabeled data might be used for unsupervised learning tasks (where the model finds patterns and structures in the data).

ML benefits from large amounts of data to achieve meaningful task accuracy. In agriculture, obtaining vast and diverse data can sometimes be challenging yet pivotal for the success of ML models. The widespread adoption of IoT technology facilitates continuous and real-time data acquisition, enabling the generation of extensive datasets over time. However, it is essential to consider that the data should be collected with sufficient quality to ensure its representativeness in the case study.

For instance, in crop management, studying the different stages of the crop is important for developing models that are accurate and applicable to real-world scenarios. Obtaining such representative datasets may take time, but it is a necessary investment for the effectiveness and reliability of ML applications in agriculture (Figure 11.1.). Furthermore, collaborative initiatives and partnerships with farmers, agricultural institutions, and research organizations can contribute to the pooling of data resources (Doe, 2024).

a. Supervised Learning: It can be used for classification tasks such as crop disease identification based on images, predicting crop yields, and recommending suitable agricultural practices.
b. Unsupervised Learning: To help with targeted extension services, farmers might be grouped according to their requirements or behavior using clustering techniques.

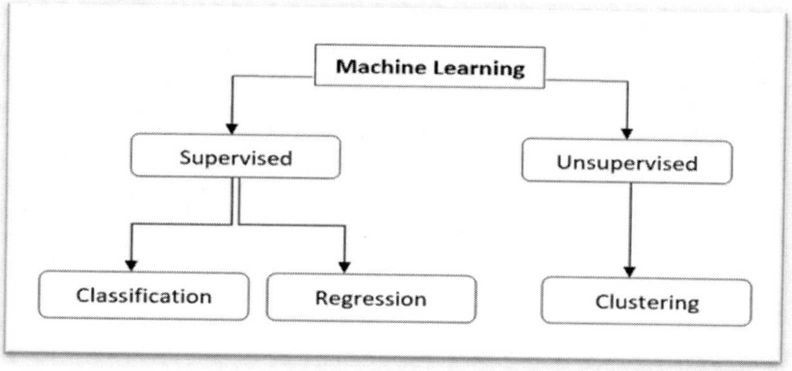


**Figure 11.1.** Machine Learning (ML).

## 11.3.2. Natural Language Processing (NLP)

Chatbots and virtual assistants powered by NLP can provide personalized advice to farmers, answering queries in natural language and disseminating information on agricultural best practices, weather forecasts, market prices, etc.

Sentiment analysis of farmer feedback can help extension workers understand the concerns and needs of the agricultural community, enabling them to tailor their services accordingly (Doe, 2022).

## 11.3.3. Computer Vision

AI-powered image analysis can assist in crop monitoring, pest and disease detection, and weed identification, facilitating early intervention and precision agriculture practices.

Remote sensing data processed through computer vision techniques can provide valuable insights into crop health, soil moisture levels, and environmental conditions, aiding in decision-making and advisory services (Garcia et al., 2022).

### 11.3.4. Expert Systems

Knowledge-based systems can encapsulate the expertise of agricultural specialists and extension workers, providing on-demand advice and recommendations to farmers on various aspects such as crop selection, pest management, and soil fertility improvement.

Rule-based systems can automate routine tasks in extension services, such as diagnosing common crop diseases based on symptoms reported by farmers (Doe, 2022).

### 11.3.5. Robotics and Automation

Autonomous drones equipped with AI algorithms can survey agricultural fields, monitor crop health, and assess land conditions, providing valuable data for extension workers and farmers to optimize agricultural practices.

Agricultural robots can automate tasks such as planting, weeding, and harvesting, reducing labor requirements and increasing productivity, thereby influencing extension service recommendations (Doe, 2022).

## 11.4. Big Data

Big data serves as the foundation for developing, testing, and optimizing AI algorithms. AI and big data go hand in hand. While large data would remain untapped without AI, AI solutions cannot be successfully implemented. AI systems are faster more and more accurate than humans at spotting patterns in observed data, creating explanatory models, and making predictions. The next great revolution in the collection and analysis of large data and the application of analysis to policymaking and improving outcomes may occur from adopting AI solutions for big data analytics. Significant data platform advancements that enable ML pattern recognition across n across numerous layers of input and granular real-time information collecting could improve the accessibility, scalability, and fine-tuning of data. The feedback loop between learning, policy formulation, investment, and result monitoring can be shortened by having access to real-time information, which enables development actors to implement change more swiftly and broadly (Doe, 2024).

While the private sector currently uses big data primarily, governments all over the world are working hard to integrate big data analytics into their operations in an attempt to improve statistical capacity for evidence-based decision-making and to address important issues like food insecurity, disease outbreaks, climate change, etc. For the benefit of the underserved and frequently inaccessible low-income communities, policymakers are starting to see the possibility of converting these vast data torrents into usable information that can be used to detect needs, deliver services, and predict and prevent disasters. Because of this, regulatory and policy programming can be experimental and flexible, allowing it to respond and adjust to complex and dynamic situations by using real-time data.

## 11.5. Types of Big Data

Three categories apply to big data as per Table 11.1: (A) structured, (B) semi-structured, and (C) unstructured.

**Table 11.1.** Types of big data and their characteristics

| S.No. | Types of Big Data | Characteristics |
|---|---|---|
| 1. | Structured Data | • Organized in fixed formats, often in rows and columns (e.g., databases, spreadsheets).<br>• Easy to store, query, and analyze using traditional tools. |
| 2. | Semi-structured Data | • Partially organized, does not conform to rigid structures (e.g., JSON, XML files).<br>• Contains tags or markers to separate data elements.<br>• Easier to analyze than unstructured data but not as straightforward as structured data. |
| 3. | Unstructured Data | • Not organized in a predefined manner (e.g., text files, videos, images, social media posts).<br>• Difficult to store, search, and analyze without specialized tools. |

## 11.6. Role of AI and Big Data in Extension Services

Extension services are essential for providing farmers with access to agricultural technology, knowledge, and practices that will help them increase their livelihoods, productivity, and sustainability. AI (Artificial Intelligence) and Big Data are increasingly becoming integral components of extension services, revolutionizing the way information is collected, analyzed and delivered to farmers.

In the past, agricultural extension services provided farmers with knowledge, instruction, and direction by relying on human skills and field-based contacts. These services have proved helpful, but there are often limitations to their scalability, reach, and timeliness. By utilizing data, algorithms, and intelligent systems, AI, which encompasses automation, machine learning, and data analytic scan significantly improves the efficacy of agricultural extension (Smith et al., 2020).

### 11.6.1. Data Collection and Analysis

Big Data and artificial intelligence (AI) are transforming extension services by offering robust instruments for gathering, analyzing, and making agricultural decisions. Using sophisticated algorithms and data analytics, AI can interpret enormous amounts of farm data from various sources, including satellites, drones, and sensors. This allows for real-time monitoring of crop health, pest infestations, and weather trends.

In today's data-driven environment, the value of AI in data collecting and analysis cannot be emphasized. AI technologies are crucial in improving the scalability, accuracy, and efficiency of data collection and analysis procedures in various fields. AI makes it possible to analyze in real-time, allowing businesses to react quickly to changing circumstances and make informed decisions in ever-changing situations. Artificial All things considered, artificial intelligence (AI) is a crucial component in releasing the potential of data by optimizing data-gathering procedures, revealing insightful information, and enabling businesses to make more informed decisions based on evidence from data.

Big data analysis is the process of looking through large, complicated datasets to find insightful information that can be put to use. Organizations in a wide range of industries are currently faced with massive amounts of data coming from a variety of sources, including sensors, social media, transaction records, and many more, in the digital age. Big data analysis uses advanced technology and procedures to clean, pre-process, analyze, and interpret data in ways that are not always visible with conventional analysis techniques. Organizations may make data-driven choices, streamline operations, spot innovation opportunities, and obtain a competitive edge in their particular industries by utilizing the potential of extensive data analysis. Applications for big data analysis range from forecasting market trends to optimizing consumer experiences and increasing operational effectiveness.

#### *11.6.1.1. Sensor Technologies and Internet of Things in Agriculture*

Sensor technology is revolutionizing agriculture by providing farmers with real-time data on various environmental and crop conditions, enabling precise decision-making and resource management. These sensors can monitor soil moisture levels, temperature, humidity, pH levels, nutrient content and even plant health indicators such as leaf color and biomass. By deploying sensors throughout their fields, farmers can gather detailed information about the status of their crops and soil, allowing them to optimize irrigation schedules, apply fertilizers and pesticides more efficiently and detect potential problems such as diseases or pests early on. Furthermore, sensor data can be integrated with other technologies, such as drones and satellite imagery to provide a comprehensive view of farm operations and facilitate precision agriculture practices.

In agriculture, the Internet of Things (IoT) operates by connecting various devices, sensors and equipment across the farm to gather real-time data and facilitate informed decision-making. IoT devices such as soil moisture sensors, weather stations, drones and automated machinery are deployed throughout the farm to monitor environmental conditions, crop health and equipment performance. These devices collect data on parameters such as soil moisture levels, temperature, humidity, rainfall and pest presence. The collected data is transmitted wirelessly to a central hub or cloud-based platform using communication technologies such as Wi-Fi, cellular networks or LPWAN. Once the data is collected, it is analyzed using advanced analytics techniques to provide insights into crop growth patterns, irrigation requirements, pest outbreaks and other factors affecting farm productivity. Farmers can then use this information to optimize their operations, improve resource efficiency and maximize yields. Overall, IoT technology in agriculture enhances productivity, sustainability and profitability by enabling data-driven decision-making, automation and remote management of farm operations.

Thanks to the Internet of Things and smart sensors, traditional agricultural methods are being replaced by intelligent farming, characterized by more production. When IoT-enabled technology is used, it is also simpler to assess crop quality, soil health, soil erosion, fertilizer demands, and soil fertility in agriculture (Sishodia et al., 2020).

### *11.6.1.2. Remote Sensing and Satellite Imagery*

Remote sensing involves acquiring and interpreting information about an object, area, or phenomenon without physical contact. In agriculture, remote sensing is crucial in monitoring crop health, assessing environmental conditions, and managing natural resources. Remote sensing techniques utilize techniques from various platforms such as satellites, drones, aircraft, and ground-based sensors to capture imagery and data across different spectral bands. These images are then processed and analyzed to extract valuable insights about vegetation health, soil moisture levels, crop stress, pest infestations, and other parameters. By providing farmers and land managers with timely and accurate information, remote sensing enables them to make informed crop decisions regarding crops. Management practices, reacquiring

Satellite imagery currently plays a crucial role in modern agriculture by providing a bird's-eye view of vast agricultural landscapes and enabling farmers, researchers and policymakers to monitor crop health, assess environmental conditions and make informed decisions. These high-resolution images, captured by satellites orbiting the Earth, offer valuable insights into various aspects of agriculture, including crop growth patterns, soil moisture levels, pest infestations and vegetation health. By analyzing satellite imagery, farmers can identify areas of stress or underperformance within their fields, allowing them to take timely action to mitigate potential losses and optimize yields. Furthermore, satellite imagery aids in precision agriculture by enabling the delineation of management zones and the implementation of site-specific farming practices, such as variable rate application of inputs and targeted irrigation.

India's agriculture industry has not yet fully embraced cutting-edge technologies. However, it is now essential to utilize modern technologies to provide sensible and effective farm management given the issues posed by rapidly degrading land, water scarcity, and climate change. While remote sensing methods have many benefits, they are not without restrictions. For example, poor access to remote sensing data on cloudy days, particularly optical data, and challenges with crop monitoring in real time (Pokhariyal et al., 2023).

#### *11.6.1.3. Social Media and Web Scraping for Data Collection*

Social media and web scraping are powerful tools for data collection, offering access to vast amounts of publicly available information across various online platforms. Social media platforms like Twitter, Facebook and Instagram serve as rich sources of real-time data, providing insights into public opinion, consumer preferences and emerging trends. Through social media monitoring, organizations can gather valuable information about their target audience, competitors and market dynamics, enabling them to tailor their marketing strategies, improve customer engagement and enhance brand reputation. Additionally, web scraping involves extracting data from websites using automated bots or software tools. This technique allows businesses to collect information from diverse sources such as news articles, product reviews and industry reports.

The two key actors in the social media effects study are the social media platform and the user, defined as any individual participant who uses a platform's capabilities to generate digital trail data. Media impacts are the result of interactions between users and platforms. Due to their selectivity and transnationality, which are likely to influence effects more specifically for a single user than for an aggregate group of users, social media effects theory places a greater emphasis on the user than mass media effects theory (Valkenburg, 2022).

### 11.6.2. Predictive Analytics for Decision Support

Predictive analytics serves as a powerful tool for decision support by leveraging historical data, statistical algorithms and machine learning techniques to forecast future outcomes and trends. By analyzing patterns and relationships within large datasets, predictive analytics enables organizations to anticipate potential scenarios, identify risks and capitalize on opportunities before they arise. In various industries, including finance, marketing, healthcare and supply chain management, predictive analytics helps decision-makers make informed choices regarding resource allocation, risk mitigation and strategic planning. For example, in finance, predictive models can forecast market trends, identify investment opportunities and assess credit risks. In healthcare, predictive analytics can predict patient outcomes, detect diseases early and optimize treatment plans. In agriculture, it plays a vital role, this can resolve many problems, and losses can be reduced to a certain extent. By harnessing the power of predictive analytics, organizations can gain a competitive edge, improve operational efficiency and enhance decision-making processes across all levels of the organization.

To achieve agriculture 4.0 goals more comprehensively, future research must also employ optimization and simulation techniques to minimize food waste, improve food security, and optimize agricultural resources to boost precision and smart agriculture productivity. This is because rapid population expansion is anticipated over the next years, raising demand for agri-food (Yousaf et al., 2023).

#### *11.6.2.1. Crop Yield Forecasting*

Crop yield forecasting is the process of predicting the amount of agricultural produce that will be harvested from a specific area or region over a given period, typically a growing season. This forecasting is crucial for farmers, policymakers, traders and food security organizations to plan and make informed decisions regarding production, distribution, pricing, and risk management. Various methods are used for crop yield forecasting, including statistical models, machine learning algorithms, remote sensing techniques and crop simulation models. These methods utilize a combination of historical data, environmental factors (*viz.* soil quality, weather conditions and pest outbreaks) and satellite imagery to estimate crop yields accurately. By providing early insights into potential crop production levels, yield forecasting helps stakeholders mitigate market volatility, optimize resource allocation and respond effectively to food security challenges and climate variability.

Agriculture plays a major role in the world economy. As the world's population grows, resolving concerns about food security and reducing the effects of climate change requires an understanding of global agricultural production. Forecasting crop yields is a major issue in agriculture. Pesticides and weather (rain, temperature, etc.) have a big influence on agricultural productivity. Decision-making related to agricultural risk management and yield forecasting depends on accurate historical crop yield data (Kavita and Mathur, 2020).

#### *11.6.2.2. Pest and Disease Outbreak Prediction*

Pest and disease outbreak prediction is a critical aspect of agricultural management, utilizing various data-driven techniques to anticipate and mitigate the impact of harmful pests and diseases on crops. By analyzing environmental factors, historical data and biological indicators, predictive models can forecast the probability of pest infestations and disease outbreaks, empowering farmers to proactively employ control tactics and preventive measures

These models may incorporate factors such as weather patterns, crop phenology, pest life cycles and geographical data to identify high-risk areas and vulnerable crops. Furthermore, early detection and response to pest and disease threats contribute to improved food security, increased resilience of agricultural systems and enhanced economic viability for farmers.

The major challenge in the agriculture sector was the reduction in crop productivity and thus income of farmers due to disease or pest outbreaks. A prediction system mitigates these problems by giving farmers an early warning, which helps them implement effective management strategies to reduce the risks of outbreak. Due to the nonlinear nature of sequences of ecological processes, accurate prediction of disease outbreaks or pest attacks is a very challenging task. For example, weather conditions, soil peculiarities, and plant characteristics influence the outbreak of diseases or pests. Pathogens at canopy or field level do not constitute a severe outbreak unless favorable meteorological conditions are met for the pathogen's growth and dissemination. The weather can alter a process's reaction rate without being a component of the system (Singh et al., 2022).

#### *11.6.2.3. Weather Pattern Analysis*

Weather pattern analysis is used to analyze historical weather data to find trends, patterns, and anomalies in the meteorological conditions across a given period and region. This analysis includes several variables that were gathered from weather stations, satellites, and other monitoring equipment, including temperature, precipitation, humidity, wind speed, and air pressure.

Meteorologists, climatologists, and academics can learn more about seasonal trends, long-term climate change implications, and climate variability by examining past weather patterns. In addition, weather pattern analysis is essential for decision-making, risk assessment, and resource planning in a number of industries, including agriculture, transportation, energy, and disaster management.

Weather forecast data is essential for making quick modifications to daily agricultural operations, which reduces losses due to bad weather and raises the quantity and quality of agricultural output. The lives of farmers in areas with significant inter-annual rainfall fluctuation are enhanced by seasonal forecasts. Seasonal forecasts heavily influence government economic policies; for example, if a significant drought is anticipated, monetary policy could be eased to meet growth targets (Shankar et al., 2019).

### 11.6.3. Personalized Recommendations and Advisory Services

Personalized recommendations and advisory services powered by AI are revolutionizing agriculture by providing tailored insights and guidance to farmers. Leveraging vast amounts of data, AI algorithms analyze factors such as soil composition, weather patterns, crop health and historical yields to generate customized recommendations for individual farms. These recommendations cover various aspects of agricultural management, including crop selection, planting schedules, irrigation techniques, pest and disease management, fertilizer application and harvesting practices. By considering specific conditions and constraints unique to each farm, AI-powered consulting services reduce environmental impact, increase productivity, and optimize resource usage.

The unpredictability of weather poses a significant risk to agricultural production. An agricultural advisory that incorporates weather forecasting can help make decisions at the field level ahead of time and mitigate associated risks. In the current context of information digitization, expanding the Agromet Field Unit network, using advanced tools and methodologies, adding more observatories and implementing customized decision support are all very effective ways to enhance weather-smart agriculture (Singh et al., 2023).

#### *11.6.3.1. Market Intelligence for Pricing and Trading Decisions*

Market intelligence serves as the bedrock for informed pricing and trading decisions in financial markets. By harnessing data analytics and AI technologies, market participants can glean a dig into market trends, sentiment shifts and competitive dynamics. These insights enable traders and investors to anticipate price movements, identify opportunities and manage risks effectively. Through real-time news, social media and market data monitoring, AI-driven market intelligence platforms provide actionable recommendations tailored to specific trading objectives and risk profiles. Whether it is optimizing entry and exit points, adjusting portfolio allocations, or executing complex trading strategies, market intelligence empowers decision-makers to navigate the intricacies of financial markets with greater precision and confidence. Ultimately, leveraging market intelligence enhances decision-making processes, maximizes returns and strengthens competitive advantages in an ever-evolving trading landscape.

Due to developments in the economy and the needs of farm households beyond meeting their food needs, agriculture has evolved from subsistence to a market-oriented sector. Instead of being production-oriented, agriculture has changed to become one that is driven by the market. Market-led research and market extension are two of the eyes of market-led agriculture, which requires an entirely different strategy (Angles and Chinnadura, 2018).

#### *11.6.3.2. Financial Planning and Risk Management Support*

Financial planning and risk management support are critical components of sound financial decision-making for individuals, businesses, and organizations. Financial planning involves setting goals, assessing resources and developing strategies to achieve desired outcomes such as wealth accumulation, retirement planning, or funding major expenses like education or housing. On the other hand, risk management entails identifying, analyzing and mitigating potential risks that could impact financial goals, including market volatility, economic downturns, regulatory changes, or unexpected events like illness or natural disasters. By integrating financial planning and risk management, individuals and entities can confidently navigate uncertainties, secure their financial future, and achieve long-term prosperity.

Agribusiness product pledges, price fluctuations and the degradation of agricultural product pledges are the primary factors influencing risks. It is possible to control the risk associated with certain agricultural pledges by considering how simple they are to realize and sell. There are two ways to assess the risk associated with changes in the market price of agricultural products used as collateral. First and foremost, we must closely monitor the collateral's price elasticity. In general, collateral that is more appropriate for use as collateral for agricultural products has lower price elasticity (Yang et al., 2022).

## 11.7. Case Studies and Examples

India is currently working very much on this issue. Here are some great efforts related to artificial intelligence and extensive data analysis.

### 11.7.1. AI-Driven Precision for Agriculture Initiatives

The World Economic Forum's Artificial Intelligence for Agriculture Innovation (AI4AI) project addresses these issues and assists India's agricultural revolution by promoting the use of artificial intelligence (AI) and related technologies for agricultural breakthroughs. This effort, which brings together government, academic, and commercial authorities to create and execute innovative ideas for the agriculture sector, is being led by the Centre for the Fourth Industrial Revolution (C4IR) India.

The 'Saagu Baagu' pilot, developed in collaboration with the Telangana state government and implemented by Digital Green in the Khammam district, is one of the most successful uses of AI4AI. It is supported by the Bill and Melinda Gates Foundation. More than 7,000 growers now have a better value chain for chilies thanks to the project. The Telangana state government, along with other helpful laws, established the nation's first framework for agri data management and interchange, which has enabled this development.

In its inaugural year, Saagu Baagu has shown impressive outcomes. Increases in quality reported by program participants led to a 21% increase in chili yields per acre, a 9% decrease in pesticide use, a 5% decrease in fertilizer use, and an 8% increase in unit prices. These changes have enhanced farmers' earnings by about INR 66,000 (around 800 USD) per acre per crop cycle, nearly doubling their income. These numbers demonstrate Saagu Baagu's efficacy and how it supports productive and sustainable farming methods.

### 11.7.2. Artificial Intelligence for Irrigation

Eighty-five percent of the freshwater resources on Earth are used for agriculture sector. Due to rising food prices and population expansion, this percentage is rising quickly. Therefore, we need to create more efficient methods to ensure that water resources are used for irrigation in the right way. Automatic irrigation scheduling techniques took the place of manual watering, which relied on soil moisture content measurements. When using an autonomous watering system, consideration was given to the evapotranspiration of plants, which is influenced by many meteorological factors including humidity, wind speed, and sun radiation, as well as crop features like development stage, plant density, soil attributes, and pests.

This method used several sensors, each serving a different function. For example, the temperature sensor records temperature, the molecular sensor improves crop development, the soil moisture sensor measures soil moisture content, the pressure regulator sensor keeps pressure levels constant, and the digital camera setup sensor finds soil moisture content. Each of these devices' outputs is transformed into a digital signal before being transferred via a wireless network, like a hotspot or Zigbee network, to a multiplexer.

The irrigation robot's performance is significantly influenced by the placement of its sensors. A single sensor can be used to regulate the watering of zones across several fields. Moreover, different zones can be irrigated using different sensors. In the first scenario, when a single sensor irrigates many zones, the sensor is placed in the driest zone, or the zone that needs the most watering, to guarantee that the entire field receives enough irrigation.

For the plants to absorb water from the sensors, they should be placed in their root zone (ensuring that there are no air gaps surrounding the sensors). This will ensure that there will be enough water available for the crops. Eventually, the sensor and SMS controller need to be connected. The controller will take over the procedure when the sensor responds. Establishing the soil water threshold is necessary after this connection has been made. The sensor's burial place is then covered with water, and the area is left unattended for a day. As previously indicated, the sensor's threshold for scheduled watering is now determined by the water content (Talaviya et al., 2020).

### 11.7.3. Artificial Intelligence for Pest Management

FarmSense, a platform for growers to detect pests, is a prime example. Insects and environmental data are collected by sensors in the field, and AI processes them once they reach the cloud. Growing knowledge about the crops, weather, and topography specific to each grower is acquired by the AI. The more effective focused strategies it can suggest.

An intelligent monitoring system is created based on gathered from examining the traits of several pests of important crops like wheat, rice, maize, potatoes, tomatoes, and numerous other vegetables. Based on data processing using MATLAB (an image processing program), this technique allows a farmer to protect their crop while strategically avoiding helpful insects. These insects need to be controlled in an environmentally and human-health-safe way.

The majority of farmers apply excessive amounts of pesticides to their fields because they are unaware of the amount required to manage the insect pests. Extra spraying eliminates beneficial and detrimental insects from the crop, and the chemicals seep into the ground, altering the crop's nutritional value and the water we drink. The precise amount of insecticide needed to control insects can be determined if their behavior is easily observed.

Farmers can also improve crop quality and yield by using various sensors in the field. The feasibility of employing wireless sensor networking for automated insect pest monitoring in agricultural settings. Our primary objective is to develop a control system and pest monitoring that leverages wireless sensor networks (WSN) to identify insect infestations early on and alert farmers to the area, thus minimizing the labor-intensive task of thoroughly scanning the entire field. Based on sensory data collected in the field, an automatic detection and classification method is utilized to identify the pest. In the future, hardware for the recommended sensor circuits for agriculture can be produced. Moreover, hardware circuitry combined with GPS and GSM technology may pinpoint the exact position of the infestation. The farmer can then take the necessary steps to protect the crop and increase productivity by receiving an emergency alert identifying the location of the infestation in their field (Singh et al., 2022).

## 11.8. Opportunities and Challenges of AI and Big Data Analysis in Extension Services

AI and big data analysis present numerous opportunities to enhance the efficiency and effectiveness of extension services, which are crucial for disseminating knowledge and technologies, particularly in agriculture, education, and public health. Here are some key opportunities:

- *Precision Agriculture*
  - *Crop Monitoring*: Crop health, soil conditions, and insect infestations may all be tracked in real-time by AI-powered drones and sensors, enabling prompt interventions.
  - *Yield Prediction*: By using historical data, weather patterns, and present growing conditions, big data analysis can forecast agricultural yields, which helps with resource allocation and planning.

- *Personalized Advisory Services*
  - *Tailored Recommendations*: AI can analyze individual farm data to provide customized advice on crop management, pest control, and irrigation, enhancing productivity and sustainability.
  - *Decision Support Systems*: Machine learning models that provide insights based on a combination of historical data, present situations, and predictive analytics can assist farmers in making well-informed decisions.
- *Enhanced Data Collection and Analysis*
  - *Real-time Data Gathering*: IoT devices and remote sensing technologies can collect vast amounts of data from the field, which can be processed and analyzed to provide actionable insights.
  - *Big Data Integration*: Forecasting and decision-making are enhanced when data from multiple sources (*viz.* the soil, weather and movements of the market) are combined to allow for in-depth study.
- *Early Warning Systems*
  - *Pest and Disease Outbreaks*: AI algorithms can detect patterns and predict pest or disease outbreaks, enabling proactive measures to prevent widespread damage.
  - *Weather Forecasting*: Accurate weather forecasts from advanced models can help farmers manage their operations and reduce the risks associated with unfavorable weather.
- *Capacity Building and Training*
  - *E-learning Platforms*: AI-driven e-learning platforms can offer personalized training programs for farmers, extension workers, and other stakeholders, enhancing their knowledge and skills.
  - *Virtual Assistance*: Chatbots are virtual personal assistants with AI features that AI drives can provide users with instant support and information to help them solve problems and make decisions.
- *Market Intelligence and Supply Chain Optimization*
  - *Market Trends Analysis*: Big data analysis can identify market trends and consumer preferences, helping farmers align their production with market demands.
  - *Supply Chain Efficiency*: AI can optimize supply chain logistics, reduce waste, and improve the efficiency of getting products from farm to market.

- *Policy and Planning*
  - *Data-driven Policy Making*: Governments and organizations can use big data analytics to develop evidence-based policies and strategies that address the needs of farmers and other stakeholders.
  - *Resource Allocation*: AI can help prioritize resource allocation by identifying the areas and interventions that will have the greatest impact.
- *Risk Management*
  - *Insurance Models*: AI can develop more accurate risk assessment models for crop insurance, providing better coverage and reducing the financial impact of adverse events on farmers.
  - *Disaster Response:* AI can help with the creation of quick response plans for natural disasters, guaranteeing aid and support for recovery in a timely manner.

The integration of AI and big data analysis into extension services holds significant potential to transform traditional practices, making them more efficient, responsive, and sustainable. By leveraging these technologies, extension services can better support farmers, improve productivity, enhance resilience, and contribute to the overall development of rural communities.

While AI and big data analysis offer significant opportunities for enhancing extension services, several challenges must be addressed to realize their full potential. These challenges span technical, social, and economic dimensions.

- *Data Quality and Availability*
  - Incomplete or Inaccurate Data: Ensuring data's accuracy, completeness, and reliability is essential. Inadequate data quality can result in inaccurate conclusions and suggestions.
  - Data Silos: The fragmentation of data across various systems and organizations poses a challenge to the comprehensive integration and analysis of data.
- *Technical Infrastructure*
  - Limited Connectivity: Rural areas often lack reliable internet access and technological infrastructure, hindering the deployment and use of AI and big data tools.

  - Hardware Constraints: High costs and maintenance requirements of advanced hardware (e.g. sensors, drones, and servers) can be prohibitive for many extension services.
- *Skill Gaps*
  - Lack of Expertise: The expertise needed to create, deploy, and manage AI and big data solutions in extension systems is frequently lacking.
  - Training Needs: To properly use and benefit from these technologies, extension workers and farmers must engage in ongoing training and capacity building.
- *Privacy and Security*
  - Data Privacy Concerns: Data collection and analysis creates questions regarding sensitive personal information security and privacy.
  - Cyber security Threats: Increased digitization and connectivity expose systems to cyber security risks, requiring robust measures to protect data and infrastructure.
- *Cost and Sustainability*
  - High Initial Investment: The costs associated with developing and deploying AI and big data solutions can be high, posing a barrier for many organizations.
  - Long-term Sustainability: Ensuring the long-term sustainability of these technologies involves continuous funding, maintenance, and upgrades.
- *Regulatory and Ethical Challenges*
  - Regulatory Compliance: Navigating the regulatory landscape for data collection, storage, and usage can be complex and time-consuming.
  - Ethical Considerations: Ensuring ethical use of AI and big data, including fairness, accountability, and transparency, is critical to maintaining trust and legitimacy.
- *Evaluation and Impact Assessment*
  - Measuring Impact: Evaluating the effectiveness and impact of AI and big data interventions in extension systems can be challenging, requiring robust methodologies and metrics.
  - Feedback Mechanisms: Establishing effective feedback mechanisms to continuously improve AI models and data analytics processes based on user experiences and outcomes.

To tackle these obstacles, a multifaceted strategy comprising cooperation between local communities, businesses, government agencies, and research institutions is needed. To effectively utilize the potential of artificial intelligence (AI) and big data analysis in extension systems, investments in infrastructure, education, and legal frameworks will be necessary, along with initiatives to assure data quality, privacy, and ethical use.

## 11.9. Recommendations for Leveraging Technologies to Enhance Decision-Making in Extension Services

In extension services, particularly in agriculture, education, and public health, integrating artificial intelligence (AI) and big data presents transformative potential. These are the several key recommendations for leveraging technologies to enhance decision-making processes:

### 11.9.1. Data Collection and Integration

Effective data collection and integration are pivotal for enhancing decision making processes through AI and big data analysis. Organizations can gather real-time, high-resolution data across various domains by employing advanced data collection techniques such as IoT devices, drones, and remote sensors. Once integrated from diverse sources like satellite imagery, local surveys, and historical records, this data creates a rich, comprehensive dataset. Such integration allows AI algorithms to process and analyze the data more effectively, uncovering hidden patterns and insights. These insights enable predictive analytics and informed decision-making, whether it is optimizing crop yields in agriculture, identifying disease outbreaks in public health, or personalizing learning experiences in education. By ensuring data quality and fostering robust data governance, organizations can maximize the potential of AI and big data, leading to smarter, data-driven decisions and improved outcomes.

### 11.9.2. Data Management and Storage

Robust data management and storage are essential for leveraging AI and big data analysis to enhance decision-making processes. By investing in scalable, secure data storage solutions, such as cloud-based systems, organizations can efficiently handle vast volumes of data generated from multiple sources. Good data management techniques guarantee the dataset's correctness and dependability. These procedures include frequent data cleansing, validation, and integration. Stringent data governance policies guarantee data privacy and security, which are crucial for maintaining stakeholder trust. With well-organized and accessible data, AI algorithms can perform comprehensive analyses, yielding valuable insights that inform strategic decisions. This structured data management approach facilitates seamless data retrieval and processing and enhances the overall efficacy of AI-driven analytics, leading to more informed, timely, and impactful decision-making across various sectors.

### 11.9.3. AI-Driven Data Analysis

Decision-making is being revolutionized by AI-driven data analysis, which makes it possible to extract significant insights from large and intricate datasets. By deploying advanced machine learning algorithms, organizations can analyze patterns and trends that are otherwise imperceptible through traditional methods. AI models, trained on diverse datasets, can predict future outcomes, identify anomalies, and provide actionable recommendations with high accuracy. This predictive capability is invaluable across sectors: in agriculture, it can forecast crop yields and optimize resource allocation; in healthcare, it can predict disease outbreaks and personalize treatments; and in business, it can anticipate market trends and consumer behavior. By continuously learning from new data, AI systems enhance their analytical precision over time, facilitating dynamic and informed decision-making. This transformative power of AI-driven data analysis enables organizations to make proactive, data-backed decisions, driving efficiency, innovation, and competitive advantage.

### 11.9.4. Decision Support Systems (DSS)

DSS powered by AI and big data analysis are essential tools for enhancing decision-making processes across various industries. These systems integrate vast amounts of data from multiple sources and apply advanced AI algorithms to generate comprehensive, real-time insights. By presenting this data through user-friendly interfaces and visualization tools such as dashboards and interactive charts, DSS makes complex information accessible and understandable for decision-makers. This makes it possible to make decisions more quickly and accurately by using data-driven information instead of gut feeling or insufficient knowledge.

In agriculture, a DSS can help farmers optimize planting schedules and resource use; in healthcare, it can aid doctors in diagnosing and treating patients more effectively; and in business, it can support strategic planning and operational efficiency. By harnessing the analytical power of AI and the expansive reach of big data, decision support systems enable organizations to anticipate challenges, seize opportunities, and achieve their goals with greater precision and confidence.

### 11.9.5. Capacity Building and Training

Capacity building and training are crucial for empowering stakeholders to utilize AI and big data analysis in decision-making effectively. Investing in comprehensive training programs ensures that extension workers, managers, and other key personnel develop essential skills in data literacy, AI tools, and analytical techniques. These programs should cover the fundamentals of data management, the interpretation of AI outputs, and the application of insights to real-world scenarios. By fostering a culture of continuous learning through workshops, online courses, and seminars, organizations can keep their teams updated on the latest advancements in AI and big data technologies. This ongoing education enhances individual competencies and promotes a collaborative, innovative environment where data-driven decision-making becomes the norm. Ultimately, well-trained personnel can leverage AI and big data more effectively, leading to more informed, strategic decisions that drive organizational success and better serve communities.

### 11.9.6. Ethical Considerations and Equity

Ethical considerations and equity are paramount in utilizing AI and big data analysis for decision-making, ensuring that these technologies benefit all stakeholders fairly and responsibly. Developing and adhering to ethical guidelines helps prevent biases in AI models, safeguarding against discrimination and ensuring transparency and accountability in data usage. These guidelines should include data privacy, consent, and security principles to protect individuals' sensitive information. Promoting equity involves making AI and big data tools accessible to marginalized and underserved communities, addressing the digital divide through affordable access to technology and training. By prioritizing inclusivity, organizations can ensure that the advantages of AI-driven insights are distributed broadly, empowering all sectors of society to make informed decisions. Ultimately, embedding ethical considerations and equity into AI and big data initiatives fosters trust, enhances the credibility of decision-making processes, and contributes to more just and equitable outcomes.

### 11.9.7. Monitoring and Evaluation

Monitoring and evaluation are critical components that enhance decision-making through the effective use of AI and big data analysis. By establishing robust monitoring and evaluation frameworks, organizations can continuously assess the performance and impact of AI-driven initiatives, ensuring they meet their intended goals and adapt to evolving needs. Regular monitoring provides real-time feedback on data quality, algorithm accuracy, and the practical implementation of insights, allowing for timely adjustments and improvements. On the other hand, evaluation offers a comprehensive analysis of long-term outcomes and the overall effectiveness of AI applications, identifying best practices and areas for refinement. Utilizing AI and big data in monitoring and evaluation processes also enables data collection and analysis automation, improving efficiency and precision. This systematic approach ensures that decision-making is not only data-driven but also responsive and iterative, leading to more effective and sustainable solutions across various sectors.

## Conclusion

Currently we are living in a world of technology where our daily life routine involves many techniques for our ease artificial intelligence and big data analysis are also a part of new technologies. When we talk about agriculture, smart agriculture, agriculture 4.0, irrigation technologies, fertilizer technologies, drone technology are there. As well as agriculture, AI plays a crucial role in medical, education, construction etc. sectors. Big data analysis helps for understanding too much data in less time. It also helps extension experts understand the farmers' needs and act likewise. The extension expert can analyze the data and give the best advice to the farmers accordingly.

Artificial intelligence (AI) and big data have transformative potential in extension services, enhancing decision-making processes across agriculture, education, public health and more. By utilizing advanced data collection techniques such as IoT devices, drones, and remote sensors, organizations can gather comprehensive, real-time data from diverse sources. Effective data management and storage solutions, coupled with stringent governance policies, ensure data quality, privacy, and security, creating a reliable foundation for AI-driven analysis. Machine learning algorithms can uncover patterns and predict outcomes, while decision support systems present these insights through user-friendly interfaces, facilitating informed, data-driven decisions. Capacity building and continuous training equip stakeholders with the necessary skills to leverage these technologies effectively. Collaborative partnerships with tech companies and research institutions bring in cutting-edge expertise and resources, ensuring practical and innovative solutions. Ethical considerations and a focus on equity ensure that AI and big data benefits are distributed fairly, addressing privacy concerns and bridging the digital divide. Finally, robust monitoring and evaluation frameworks enable continuous improvement and adaptive decision-making, ensuring the sustained effectiveness and impact of AI and big data initiatives. Through these strategies, extension services can achieve greater efficiency, innovation, and positive outcomes for communities.

## References

Agarwal, N. and Jain, S. (2020) Role of Artificial Intelligence and Big Data Analytics in Indian Transportation Sector: Enhancing Informed Decision Making for Efficient

Mobility. *International Journal of Transportation Engineering and Technology*. 7(4):112-125.

Agarwal, S. and Sharma, P. (2019) Role of Artificial Intelligence and Big Data Analytics in Indian Aviation Industry: Enhancing Informed Decision Making for Airline Operations. *International Journal of Aviation Management and Operations*. 6(3):89-102.

Angles, S. and Chinnadura, M. (2018) Impact and Importance of Market Intelligence in Indian Agriculture. *Research Journal of Agricultural Sciences* 9(Special): 117-120.

Anonymous (2024) AI for Agriculture: How Indian Farmers are Harvesting Innovation. *World Economic Forum*. https://www.weforum.org/impact/ai-for-agriculture-in india/#:~:text=India%27s%20farmers%20combat%20climate%20change,through%20agritech%20and%20data%20management.

Balyan, S., Jangir, H., Tripathi, S. N., Tripathi, A., Jhang, T., & Pandey, P. (2024). Seeding a Sustainable Future: Navigating the Digital Horizon of Smart Agriculture. *Sustainability*, 16(2), 475.

Bhatnagar, A., and Rastogi, R. (2020) Leveraging Artificial Intelligence and Big Data Analytics for Informed Decision Making in Indian Healthcare Sector. *International Journal of Advanced Research in Computer Science*. 11(3):20-28.

Choudhary, R. and Singh, A. (2021) Role of Artificial Intelligence and Big Data Analytics in Indian Pharmaceutical Industry: Enabling Informed Decision Making for Drug Development. *International Journal of Pharmaceutical Sciences and Research*, 12(4):78-90.

Doe, A. (2022). The Role of Autonomous Drones and Agricultural Robots in Modern Farming. *Journal of Agricultural Technology*, 10(4), 245-257.

Doe, A. (2024). The Synergy of AI and Big Data in Transformative Analytics. *Journal of Data Science and Policy*, 18(1), 45-58.

Doe, A. (2024). The Role of Artificial Intelligence in Modern Agriculture. *International Journal of Agricultural Science and Technology*, 20(1), 45-58.

Garcia, A., & Patel, S. (2022). AI-powered Image Analysis for Precision Agriculture: A Review. *Journal of Agricultural Technology*, 8(3), 45-58.

Gupta, A. and Mishra, S. (2020) Role of Artificial Intelligence and Big Data Analytics in Indian Judicial System: Facilitating Informed Legal Decision Making. *International Journal of Law and Legal Studies*. 7(1):98-110.

Gupta, A. and Verma, S. (2021) Role of Artificial Intelligence and Big Data Analytics in Indian Hospitality Industry: Enhancing Informed Decision Making for Customer Experience Management. *International Journal of Hospitality and Tourism Management*. 9(1):56-68.

Gupta, R. and Aggarwal, S. (2021) Leveraging Artificial Intelligence and Big Data Analytics for Informed Marketing Decisions in Indian FMCG Sector. *International Journal of Marketing and Business Communication*. 9(3):78-90.

Gupta, R. and Singh, A. (2021) Harnessing Artificial Intelligence and Big Data Analytics for Informed Disaster Management: Case Studies from Indian Context. *International Journal of Disaster Management and Risk Reduction*. 8(3):112-125.

Gupta, R. and Singh, V. (2020) Harnessing Artificial Intelligence and Big Data for Informed Investment Decision Making in Indian Financial Markets. *International Journal of Management, IT and Engineering*. 10(1):127-140.

Jain, M. and Agarwal, S. (2019) Leveraging Artificial Intelligence and Big Data Analytics for Informed Talent Acquisition Decisions in Indian IT Industry. *International Journal of Human Resource Management and Research*. 9(3):56-68.

Jain, M. and Gupta, S. (2020) Leveraging Artificial Intelligence and Big Data Analytics for Informed Recruitment Decisions in Indian Healthcare Sector. *International Journal of Health Informatics and Medical Research*. 6(2):34-45.

Jain, M. and Singh, S. (2019) Role of Artificial Intelligence and Big Data Analytics in Indian Telecom Industry: Enhancing Informed Decision Making for Network Optimization. *International Journal of Telecommunications and Networking*. 6(1): 45-57.

Jain, S. and Verma, N. (2020) Leveraging Artificial Intelligence and Big Data Analytics for Informed Water Resource Management Decisions in Indian Agriculture. *International Journal of Agricultural Science and Research*. 8(2):34-45.

Javaid, M., Haleem, A., Khan, I. H., & Suman, R. (2023). Understanding the potential applications of Artificial Intelligence in Agriculture Sector. *Advanced Agrochem*, 2(1), 15-30.

Johnson, R. (2023). The Role of AI in Addressing Global Food Insecurity. *Global Food Security Journal*, 12(1), 45-58.

Kanojia, V., Subeesh, A., & Kushwaha, N. L. (2024). Artificial Intelligence and Smart Farming: An Overview. Artificial Intelligence and Smart Agriculture: Technology and Applications, 3-22.

Kapoor, S. and Kumar, A. (2019) Role of Artificial Intelligence and Big Data Analytics in Agricultural Decision Making: A Case Study of Indian Farming Community. *International Journal of Agriculture and Environmental Research*. 5(2):45-55.

Kavita and Mathur, P. (2020) Crop Yield Estimation in India Using Machine Learning. In 2020 *IEEE 5th International Conference on Computing Communication and Automation* (ICCCA), 220–224.

Kumari, S., Pandey, G. K., & Tiwari, S. (2024). Data Visualization Techniques in Smart Agriculture Implementation. In *AI Applications for Business, Medical, and Agricultural Sustainability* (pp. 122-159). *IGI Global*.

Kutyauripo, I., Rushambwa, M., & Chiwazi, L. (2023). Artificial intelligence applications in the agrifood sectors. *Journal of Agriculture and Food Research,* 11, 100502.

Mishra, A., and Singh, R. (2020) Harnessing Artificial Intelligence and Big Data Analytics for Informed Urban Planning Decisions: A Case Study of Indian Smart Cities. *International Journal of Urban Planning and Smart Cities*. 2(2):56-67.

Mishra, P., and Sharma, S. (2021) Leveraging Artificial Intelligence and Big Data Analytics for Smart Governance: Insights from Indian Municipalities. *Journal of Public Administration and Governance*. 11(1):88-100.

Mishra, S. and Sharma, R. (2020) Leveraging Artificial Intelligence and Big Data Analytics for Informed Human Resource Management Decisions in Indian Organizations. *International Journal of Human Resource Development and Management*. 8(2):78-90.

Pachiappan, K., Anitha, K., Pitchai, R., Sangeetha, S., Satyanarayana, T. V. V., & Boopathi, S. (2024). Intelligent Machines, IoT, and AI in Revolutionizing Agriculture for Water Processing. *In Handbook of Research on AI and ML for Intelligent Machines and Systems* (pp. 374-399). IGI Global.

Pandey, D. K., & Mishra, R. (2024). Towards sustainable agriculture: Harnessing AI for global food security. *Artificial Intelligence in Agriculture.*

Patel, M. and Gupta, A. (2021) Leveraging Artificial Intelligence and Big Data Analytics for Informed Waste Management Decisions: Case Studies from Indian Cities. *International Journal of Waste Management and Sustainable Development.* 9(1):56-68.

Patel, M. and Verma, S. (2021) Harnessing Artificial Intelligence and Big Data Analytics for Informed Environmental Management Decisions: Case Studies from Indian Industries. *International Journal of Environmental Science and Sustainable Development.* 8(2): 45-57.

Patel, N. and Singh, P. (2019) Harnessing Artificial Intelligence and Big Data Analytics for Informed Investment Decision Making in Indian Real Estate Sector. *International Journal of Real Estate Research and Analysis.* 6(3):89-102.

Patel, P. and Sharma, R. (2020) Role of Artificial Intelligence and Big Data in Indian Education System: Empowering Informed Decision Making. *Journal of Education and Learning.* 9(4):112-123.

Pokhariyal, S., Patel, N. R. and Govind, A. (2023) Machine Learning-Driven Remote Sensing Applications for Agriculture in India—A Systematic Review. *Agronomy,* 13, 2302.

Reddy, K. and Kumar, S. (2019) Leveraging Artificial Intelligence and Big Data Analytics for Informed Risk Management in Indian Banking Sector. *International Journal of Banking, Risk and Insurance.* 7(1):34-45.

Sankar, T., Arul, S., Sudarmanian, N. S. and Dheebakaran, G. A. (2019) Weather Forecasting in Agriculture. *Research Trends in Agriculture Science.* Pp. 57-81.

Sharma, A. and Kumar, R. (2019) Role of Artificial Intelligence and Big Data Analytics in Indian Financial Inclusion: Enabling Informed Banking Decisions for Rural Population. *International Journal of Financial Inclusion and Social Change.* 7(1):45-57.

Sharma, P. and Singh, A. (2020) Role of Artificial Intelligence and Big Data Analytics in Indian Tourism Industry: Enhancing Informed Decision Making for Sustainable Development. *International Journal of Tourism and Hospitality Management.* 8(2), 78-89.

Sharma, R. and Mishra, A. (2020) Harnessing Artificial Intelligence and Big Data Analytics for Informed Energy Consumption Patterns: Insights from Indian Residential Sector. *International Journal of Energy and Sustainable Development.* 7(2):78-90.

Sharma, S. and Kumar, P. (2019) Leveraging Artificial Intelligence and Big Data Analytics for Informed Policy Making: Insights from Indian Government Initiatives. *International Journal of Public Policy and Governance.* 7(3):112-125.

Singh, K. D., Kumar, A., Raja, L., Kumar, V., Kushwaha, A. K. S., Vashney, N. and Chhetri, M. (2022) An Artificial Neural Network- Based Pest Identification and

Control in Smart Agriculture Using Wireless Sensor Networks. *Journal of Food Quality*. Pp. 1-12.

Singh, K. K., Ghosh, K., Bhan, S. C., Singh, P., Vishnoi, L., Balasubramanian, R., Attri, S. D., Goroshi, S. and Singh, R. (2023) Decision Support System for Digitally Climate Informed Services to Farmers in India. *Journal of Agrometeorology*. 25(2):205-214.

Singh, M., Setiya, P., Tewari, A. K. and Nain, A. S. (2022) Comparison of Machine Learning and Regression Approaches to Forecasting Alternaria blight Epidemic of Indian Mustard. *The Pharma Innovation Journal*. 11(6):2974- 2984.

Singh, R. and Sharma, A. (2019) Leveraging Artificial Intelligence and Big Data Analytics for Informed Real Estate Investment Decisions in Indian Property Market. *International Journal of Real Estate Research and Investment*. 5(3):67-79.

Singh, S. and Kumar, R. (2021) Artificial Intelligence and Big Data Analytics: Transforming Indian Retail Industry for Informed Consumer Decision Making. *International Journal of Management Studies*. 8(2):36-45.

Sishodia, R. P., Ray, R. L. and Singh, S. K. (2020) Applications of Remote Sensing in Precision Agriculture: A Review, Remote Sensing 12:3136.

Smith, J., & Johnson, M. (2020) The Role of Artificial Intelligence in Agricultural Extension for Sustainable Development. *Journal of Agricultural Science*. 45(3):225-241.

Smith, J. (2023). The Impact of AI and Big Data on Agricultural Extension Services. *Journal of Agricultural Innovation*, 15(2), 123-135.

Talaviya, T., Shah, D., Patel, N., Yagnik, H. and Shah, M. (2020) Implementation of Artificial Intelligence in Agriculture for optimisation of Irrigation and Application of Pesticides and herbicides. *Artificial Intelligence in Agriculture*. 4:58-73.

Tiwari, S. and Gupta, A. (2021) Leveraging Artificial Intelligence and Big Data Analytics for Informed Energy Management Decisions in Indian Organizations. *International Journal of Energy and Environment*. 12(3):123-135.

Valkenburg, P. M. (2022) Theoretical Foundations of Social Media Uses and Effects. In J. Nesi, E. H. Telzer, & M. J. Prinstein (Eds.), *Handbook of Adolescent Digital Media Use and Mental Health* (1st ed, pp. 39–60). Cambridge University Press.

Verma, N. and Agrawal, A. (2019) Leveraging Big Data Analytics and Artificial Intelligence for Informed Supply Chain Decision Making in Indian Manufacturing Sector. *International Journal of Research in Engineering, Science and Management*. 2(6):89-97.

Yang, J., Tang, D., Kong, H. and Boamah, V. (2022) Research on Financial Risk Management and Control of Agricultural Products Supply Chain- A Case Study of Jiangsu Province of China. *Front. Environ. Sci*. 10:1008716.

Yang, Z., Wang, X., Cheng, Y., & Liu, G. (2021). Aspect-based capsule network with mutual attention for recommendations. *IEEE Transactions on Artificial Intelligence*, 2(3), 228-237.

Yousaf, A., Kayvanfar, V., Mazzoni, A. and Elomri, A. (2023) Artificial Intelligence-based Decision Support Systems in Smart Agriculture: Bibliometric Analysis for Operational Insights and Future Directions. *Front. Sustain. Food Syst.* 6:1053921.

# Chapter 12

# Agroecology and Traditional Farming Systems: A Holistic Approach to Sustainable Agriculture

**Smaranika Mohanty[1,*]**
**Santosh Ku. Rout[2]**
**Manoranjan Dash[3]**
**Anshuman Jena[2]**
**Soumya Shakti Dash[1]**
**Sweta Sahoo[1]**
**Swatee Pragyan[1]**
**and Dibyajyoti Behera[1]**

[1]Department of Agricultural Extension and Communication,
Faculty of Agricultural Sciences, SOADU, Bhubaneswar, Orrisa, India
[2]Faculty of Agricultural Sciences, SOADU, Bhubaneswar, Orrisa, India
[3]Information Systems and Operation Management, SOADU, Bhubaneswar, Orrisa, India

## Abstract

Sustainable agriculture has become a paramount global objective to address food security, environmental degradation, and climate change. This abstract explores integrating conventional farming techniques with agroecology as a comprehensive approach to achieving agricultural sustainability. According to ecological principles, agroecology emphasizes the interplay of ecological, social, and economic factors within farming systems. Traditional agricultural practices passed down through generations offer valuable, sustainable methods that have

* Corresponding Author's Email: smaranika1423@gmail.com

In: Global Horizons in Agricultural Extension Education
Editors: R. K. Doharey, Aman Verma, Satyapriya et al.
ISBN: 979-8-89530-446-4

evolved harmoniously with local ecosystems. Agroecology minimizes reliance on external inputs such as synthetic fertilizers and pesticides by promoting biodiversity, nutrient cycling, and natural pest control. These systems enhance resilience to climate change and contribute to agro-biodiversity conservation through the symbiotic relationship between crops, animals, and natural ecosystems. Furthermore, agroecology prioritizes community engagement, information sharing, and collaborative decision-making to promote social sustainability. Well-established farming systems with strong cultural and historical roots provide a wealth of sustainable techniques that may be modified and incorporated into contemporary farming methods. In addition to supporting nearby communities' sustainability, preserving and restoring traditional knowledge offers insightful information for improving world food systems. This abstract emphasizes the necessity of a paradigm change in agricultural techniques and the synergy between agroecology and conventional farming systems to establish resilient, sustainable, and socially just food production systems. When these strategies are combined, they can help create novel solutions that maintain cultural legacy and ecological integrity while addressing the complex issues confronting agriculture.

**Keywords:** degradation, sustainability, ecological integrity, traditional knowledge

## 12.1. Introduction

The current global population of undernourished individuals stands at approximately one billion with projections suggesting a doubling of this number by 2050, aligning with the expected global population increase to 9.1 billion (Tripathi et al., 2019; Ikrang et al., 2022). Although agriculture is the primary source of sustenance and income for over 60% of the world's population, food production rates have been unable to keep pace with population growth (Porkka et al., 2017; Odusola, 2021). Conventional agricultural methods, characterized by extensive mono-cropping, high-yield crop varieties, synthetic inputs such as fertilizers and pesticides, fossil fuel-dependent mechanization, and extensive irrigation, have failed in eliminating hunger despite boosting crop yields. This raises concerns about their long-term economic, social, and environmental sustainability. Additionally, industrial agriculture accounts for a significant 25% to 30% of global greenhouse gas emissions, exacerbating climate change and threatening future food production (Liu et al., 2020).

Annual pesticide consumption surpasses 2.6 million tons, with a market value exceeding US $25 billion (Rajbhandari, 2017; Abd-Aziz et al., 2022). Widespread pesticide usage often disrupts natural regulatory mechanisms and contributes to biodiversity loss, which otherwise supports food production. While high-yield crops and synthetic fertilizers offer temporary benefits, they have proven ineffective in averting yield declines, notably in major grain and legume-producing regions (Rajbhandari, 2017; Kuyah et al., 2021). The global population growth rate surpasses food production rates, leading to significant food shortages, persistent hunger, and malnutrition, especially in the least developed nations, with several African countries bearing the brunt. Despite agriculture being the backbone of Africa's economy and a vital livelihood for most, many countries on the continent face numerous barriers hindering agricultural productivity, including soil degradation, climate change impacts, water scarcity, post-harvest losses, and limited market access, among others (Gashu et al., 2019).

The integrative study of the ecology of the whole food system, including ecological, economic, and social dimensions, is what we refer to as agroecology. This concept will pave the way for a useful strategy that will inspire researchers, educators, and students of all ages to embrace the connection and completeness of systems. It will also emphasize the distinctive qualities of each location and solutions suitable for its limitations and available resources. The concept broadens our perspective beyond farming and field-level production methods and their direct effects on the environment. More precisely, according to Gliessman (1998), "agroecology is defined as the application of ecological concepts and principles to the design and management of sustainable agroecosystems." Farmers and scientists looking for novel approaches to boost agricultural output and sustainability while preserving an ecosystem that must persist and offer a high standard of living have found great interest in this idea. Most university courses and research programs, including those in agroecology, concentrate on the specific components of agricultural production and their immediate environmental impacts, avoiding the more difficult task of changing today's food systems.

The burgeoning global population, coupled with climate change, water scarcity, and environmental degradation, pose significant challenges to agriculture worldwide. According to the United Nations (2017), the global population is projected to swell by 83 million annually, reaching 9.8 billion by 2050. Consequently, some of the world's poorest nations are expected to experience heightened hunger levels and an increased demand for expanded food production. Moreover, as the population expands, there is a growing need for fuel and animal feed (Food and Agriculture Organization of the United Nations [F.A.O.], 2017).

Unsustainable agricultural techniques, often referred to as conventional or industrial farming systems, have been adopted in many regions worldwide to meet the rising demand for food and agricultural land. These practices include genetic modification, monocropping, heavy reliance on agrochemical inputs, and unsustainable water usage in irrigated cropping systems. Studies by Aziz et al. (2015), Bhandari (2014), and Kalia and Gosal (2011) have shown that these agricultural practices have inflicted harm on natural ecosystems, exacerbated climate change, polluted water systems, and depleted soil fertility.

Therefore, it is crucial to enhance agricultural productivity using environmentally friendly techniques that do not jeopardize human health or the environment (Altieri et al., 2015; De Schutter, 2010; De Schutter, 2014). Agroecological farming techniques, which employ ecological principles to create agricultural systems that minimize external inputs and encourage ecosystem interactions, aim to ensure food security without adverse environmental impacts (Altieri and Nicholls, 2015; Altieri, Nicholls, and Montalba, 2017).

Various strategies such as crop rotations, green manuring, composting, bush fallow systems, and shifting cultivation are employed to increase soil organic matter and nutrient content. Additionally, practices like integrating livestock into cropping systems, utilizing indigenous seeds and livestock breeds, naturally managing pests and diseases, and implementing polycultures are widely recognized methods utilized in small-scale family farms worldwide as organic farming techniques (Altieri and Toledo, 2011; Wezel et al., 2014).

Organic agriculture emerged in response to concerns about conventional agricultural practices contaminating the food supply, degrading ecosystems, and contributing to climate change using agrochemicals and greenhouse gas emissions (Morgera, Caro, and Durán, 2012).

## 12.2. Agricultural Sustainability

What is meant by “sustainable agriculture”? Over time, various phrases have been used to suggest increased sustainability in various agricultural systems. The dominant ones (pre-industrial and industrialized). They include community-based, wise-use, farm-fresh, extensive, organic, ecological, low-input, biodynamic, sustainable, eco-agricultural, and permaculture. Whether agricultural systems that use some of these concepts may be sustainable is a significant and ongoing dispute.

Systems that exhibit high sustainability make optimal use of the resources provided by nature without endangering them. Some main ideas are provided below:

a. *Utilization of Natural Processes*: High sustainability systems leverage natural processes to enhance food production, including soil regeneration, nutrient cycling, nitrogen fixation, and natural pest control.
b. *Reduction of Non-Renewable Inputs*: These systems aim to minimize the use of non-renewable inputs that pose environmental risks or threaten the health of farmers and consumers.
c. *Leveraging Farmer Knowledge*: They emphasize tapping into farmers' expertise to promote self-sufficiency and replace costly inputs with human capital.
d. *Collaborative Problem-Solving*: High sustainability systems foster collaboration among stakeholders to address common agricultural and natural resource challenges, such as pest management, watershed protection, irrigation practices, forest conservation, and access to credit.

Agricultural sustainability does not include the ideological exclusion of any technologies or methods. Technology is probably better from a sustainability perspective if it increases farmer productivity while posing no environmental risks. Systems of agriculture that prioritize these ideas have multiple uses in both economics and landscapes. Along with producing commodities for farm families and markets, they support several important public goods, including wildlife, clean water, soil carbon sequestration, flood protection, groundwater recharge, and landscape amenity value.

To maximize the benefits of nature's resources, more sustainable agriculture must adapt and implement local technology and practices. These are most likely to result from new arrangements of social capital, which include trust relationships embodied in new social organizations, new partnerships between institutions, both horizontal and vertical, and human capital consisting of skills like leadership, creativity, management, and innovation.

High social and human resource agricultural systems are better equipped to innovate when faced with unpredictability. Agricultural sustainability practices are often associated with excessive use of inputs, meaning that more acreage is needed to produce the same quantity of food, despite this being false.

All available empirical data indicates that effective agricultural sustainability programs and efforts result from adjustments to agricultural production variables (such as switching from fertilizers to legumes that fix nitrogen to pesticides to a focus on natural enemies). Nevertheless, they have also necessitated reorganizations of human capital (knowledge, labor, and managerial abilities) and social capital (ability to collaborate).

Therefore, it would be wiser to propose that sustainability entails intensification of resources, or greater utilization of already-existing resources (such as land, water, and biodiversity) and technology. Many people now associate the term "intensification" with negative connotations, such as the development of agricultural systems that substantially negatively impact the environment in industrialized nations. We must ask ourselves: "What exactly do we mean by 'type of intensification'?" Sustainable intensification refers to one that minimizes or eliminates environmental impact while utilizing the greatest technology inputs (best genotypes and best ecological management). This is accomplished by utilizing natural, social, and human capital assets.

### 12.2.1. Elements of Agro-Ecology

The 10 Elements of Agro-ecology were developed through a collaborative process involving several stakeholders. The goal was to create a framework for redesigning agricultural systems and processes that can be optimized and tailored to specific local conditions. The project's development took place from 2015 to 2019 and involved three primary phases.

The phase of gathering information involved several crucial activities, including a comprehensive examination of scientific literature and extensive discussions with agroecology experts globally. Our analysis concentrated on influential works like the papers by Altieri (1995) and Gliessman (2015), along with non-peer-reviewed literature, to delve into the varied interpretations of agroecology utilized by different stakeholders across different settings.

Apart from academic sources, we integrated insights from presentations and deliberations at the First International Symposium on Agroecology for

Food Security and Nutrition (F.A.O., 2015), as well as perspectives shared in seven F.A.O. multi-stakeholder gatherings held between 2015 and 2017 in Brasilia, Dakar, Bangkok, Kunming, La Paz, Budapest, and Tunis. These meetings fostered involvement from governmental bodies, civil society, academia, and the private sector, with a comprehensive summary provided in F.A.O. (2018c).

In total, these meetings convened more than 1,400 attendees from 170 Member Countries and about 500 organizations operating across different levels, ranging from local to international. The organizers emphasized considerations such as gender balance, geographic diversity, and a diverse array of stakeholders when selecting participants and speakers eligible for financial support.

The process of synthesis, spearheaded by specialists from the Food and Agriculture Organization (F.A.O.) alongside external agroecology experts, comprised workshop sessions, email exchanges, and iterative rounds of writing and feedback. This phase aimed to identify common themes from the information gathering stage and group together expressions of ideas in line with F.A.O.'s Common Vision for Sustainable Food and Agriculture (F.A.O., 2014).

Presentations delivered at the inaugural International Symposium on Agroecology for Food Security and Nutrition (F.A.O., 2015) underscored significant themes including recycling, efficiency, diversity, resilience, and synergies, all acknowledged as fundamental ecological components of agroecology (Tittonell, 2015).

A deeper examination of recommendations arising from regional consultations (F.A.O., 2018c) identified specific areas for future emphasis within the realm of agroecology. While these recommendations closely mirrored the initial five aspects, regional dialogues also emphasized the necessity of enhancing the social and political dimensions of agroecology.

Consequently, findings from regional discussions were categorized into five supplementary elements: cooperative knowledge creation, principles related to humanity and society, cultural and culinary traditions, responsible governance, and circular and collaborative economies. After refining both the titles and contents of these components and crafting a cohesive narrative highlighting the unity, interconnectedness, and interdependence of agroecologyagroecology and its ten facets, the framework underwent comprehensive evaluation by international experts and the F.A.O.

In 2019, the global scientific community was introduced to the emerging idea of visual storytelling at the 4th World Congress on Agroforestry held in

Montpellier, France. This introduction occurred through a keynote speech in a session titled "Agroforestry and Agroecology Opportunities and Challenges" and during an F.A.O. side-event titled "Agroforestry and the Tenure Barrier," both held as part of the congress.

The gathering convened over 1,200 representatives from 97 nations and coincided with the 10th Ecosystem Services Partnership World Conference held in Hannover, Germany, which drew over 750 participants. The presentation was part of a sectoral working agroecology group day session titled "Agroecology: managing biodiversity and soil health for the sustained provision of ecosystem services in agriculture." This session was jointly organized and hosted by F.A.O., the Helmholtz Centre for Environmental Research (U.F.Z.), and the Institute of Environmental Planning at the University of Hannover.

The framework of the 10 Elements of Agroecology was unveiled at the Second F.A.O. International Symposium on Agroecology in April 2018 (F.A.O., 2018b). Following this, during its 26th Session in October 2018, the Committee on Agriculture (COAG) officially endorsed it as a blueprint for advancing sustainable food systems and agriculture (F.A.O., 2018d). This endorsement received unanimous support from all 197 Members of the Food and Agriculture Organization of the United Nations, signifying its adoption as F.A.O.'s overarching vision for Agroecology. The approval came after a rigorous review, revision, and clearance process overseen by F.A.O.'s governing bodies, as outlined in the report of the 163rd F.A.O. Council held in December 2019 (F.A.O., 2019) (Figure 12.1).

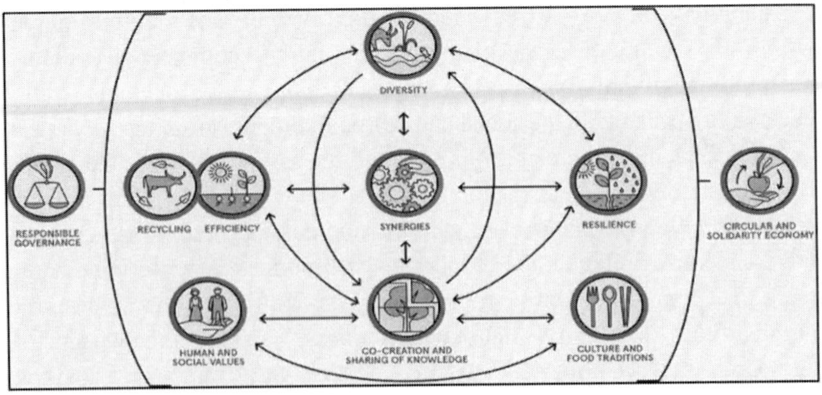


**Figure 12.1.** Facilitating shifts towards sustainable agriculture and food systems.

### 12.2.2. The Ecological Role of Biodiversity in Agro Ecosystems

The diversity found within agricultural ecosystems is pivotal in shaping their resilience and capacity to sustain crop yield, plant health, and soil quality. Experts in entomology and plant pathology underscore the significance of genetic variety in mitigating crop vulnerability to diseases and pests. Research indicates that varied cropping systems like polycultures and agroforestry arrangements encounter fewer instances of insect pests and exhibit slower disease spread, leading to diminished crop loss and enhanced yields compared to monoculture practices (Francis, 1986; Altieri, 2002).

In areas such as China, Bolivia, and Kenya, economically disadvantaged farmers depend on a range of seed types, encompassing traditional crops like rice, potatoes, and maize, for their ability to adapt and ensure survival (Swiderska, 2011). For instance, in China, cultivating diverse rice strains across various townships resulted in decreased disease occurrence and increased yields, eliminating the necessity for fungicides (Zhu et al., 2000). Moreover, incorporating tree cover within agroforestry systems can improve soil water infiltration and decrease evaporation, offering advantages to crops in regions with limited rainfall (Lin, 2007).

Intercropping offers farmers the opportunity to cultivate multiple crops concurrently, thus reducing the risks linked with monoculture agriculture (Vandermeer, 1989). Polycultures exhibit enhanced yield consistency and productivity during periods of drought when compared to monocultures. Studies conducted by Natarajan and Willey (1996) demonstrated that intercropping sorghum with millet or peanuts resulted in increased yields even under differing levels of water scarcity. Notably, the difference in productivity between monocultures and polycultures widened as water scarcity intensified.

Intensive silvo-pastoral systems (I.S.S.) represent a sustainable approach to agroforestry in cattle farming by incorporating fodder shrubs alongside trees and palms to enhance grasslands. Tactics such as electric fencing, rotational grazing, and providing water ensure high stocking densities and natural production of milk and meat. At El Hatico farm in Valle del Cauca, Colombia, implementing a five-story I.S.S. led to a 130% rise in milk output and stocking densities of 4.3 dairy cows per hectare without resorting to chemical fertilizers. Even during the severe drought of 2009, El Hatico farm maintained steady milk production owing to the resilience of its diversified agricultural system. In contrast, other regions experienced substantial weight loss and elevated mortality rates among livestock due to famine and dehydration.

Diverse agricultural systems offer ecological advantages like managing water, regulating microclimates, preserving biodiversity, and sequestering carbon, thereby bolstering resilience to climate change while also providing valuable goods and services. Intercropping enables farmers to cultivate multiple crops concurrently, reducing risks associated with monoculture farming (Vandermeer, 1989). Polycultures demonstrate enhanced stability in yields and productivity during drought conditions compared to monocultures. Studies by Natarajan and Willey (1996) illustrated that intercropping sorghum with millet or peanuts increased yields even under varying levels of water stress. Notably, the productivity disparity between monocultures and polycultures widened as water stress intensified.

### 12.2.3. Increased Environmental Sustainability Through Agricultural Ecological Practices

Historically, conventional agriculture has prioritized the production of food, feed, forage, fiber, and fuel, particularly since the advent of the Green Revolution. To meet these goals, many methods heavily rely on artificial inputs like chemical fertilizers, pesticides, and irrigation from finite water sources. However, these approaches often result in significant agricultural drawbacks, including pollution from agrochemicals, incidents of pesticide poisoning, and emissions of greenhouse gases. Moreover, they frequently disrupt critical ecological processes, such as biotic and abiotic interactions, which are essential for providing ecosystem services crucial for sustainable agriculture, such as maintaining natural soil fertility, biological pest control, water retention capacity, and resilience to extreme events (Zhang et al., 2007).

According to the framework for ecosystem services (E.S.) established by the Millennium Ecosystem Assessment (Reid et al., 2005), agricultural systems rely on "input services" (supporting and regulating E.S.) such as biological control, water purification, and nutrient cycling, alongside providing "output services" (provisioning and cultural E.S.). The objective of developing ecologically sustainable farming methods is to optimize the balance between input and output services (Zhang et al., 2007). Wezel et al. (2014a) characterized agroecological practices as agricultural approaches that prioritize significant food production while incorporating ecological processes and ecosystem services as fundamental components.

Nonetheless, Pfiffner et al. (2009) propose that simply having wildflower strips may not be adequate for pest control, as their efficacy depends on insects from (semi-)natural landscape habitats migrating to these strips. Various methods necessitate interactions between plot size and landscape to maintain ecosystem services. For example, to uphold biological pest control, Tamburini et al. (2016) discovered that conservation tillage—characterized by soil management that doesn't involve inversion and is often combined with permanent plant cover—can be successful even in less complex environments.

## 12.3. Organizing the Food System to Elevate Its Societal Importance and Ensure Economic Viability

As previously mentioned, a production system that relies on natural processes rather than inputs creates pressure on the entire food chain by increasing the diversity of products over time and location (Kremen et al., 2012a). This challenge becomes particularly significant given the wide range of products generated by agricultural systems such as fuel, feed, forage, and fiber (Delcour et al., 2014). The conventional food system, which is largely controlled by a handful of international food corporations and revolves around the supermarket model, prioritizes competitiveness in price, uniformity of products, and efficiency in logistics (Raynolds, 2004). The standardization of products has been facilitated by mechanization and the use of external chemical inputs (Marsden et al., 2006), allowing food corporations, which dominate the food system, to expand globally in order to meet demands for efficiency in logistics and competitiveness in price. This global expansion intensifies competition among farmers and boosts short-term productivity (Kremen et al., 2012a; Rosset et al., 2012).

The considerable decrease in the number of farmers and the heavy dependence on E.U. subsidies as a source of farmer income underscores the limitations of the current economic model for agriculture in the E.U. This situation highlights the importance of developing sustainable food systems and eventually integrating them into agroecology. The establishment of such systems should involve various stakeholders, including farmers, processors, distributors, consumers, scientists, and policymakers, as proposed by Francis et al. (2003). Gliessman (2011) emphasized that farmers alone cannot bring about a revolution in the entire food system. Recently, a clearer approach has emerged through a set of 13 guiding principles for sustainable food systems, as outlined by Dumont et al. (2016). These principles encompass aspects like financial self-sufficiency, environmental fairness, partnerships between consumers and producers, and physical proximity.

### 12.3.1. The Implementation of Agro Ecological Methods Among Small-Scale Farmers: Effects on Food Sovereignty

In recent times, there has been a growing worldwide interest in small-scale agriculture, driven by its potential to tackle food security challenges amidst energy, economic, and climate-related crises (Simon et al., 2020). Furthermore, the principles of agroecology align with the food sovereignty movement and provide support to small-scale farmers. According to La Via Campesina, food sovereignty encompasses the right of communities to determine their own food and agriculture systems, prioritizing the production of healthy, culturally appropriate food in a sustainable manner (Huambachano, 2019). This concept offers a pathway toward achieving food security and has resulted in increased focus on agroecological-based agricultural methods and food sovereignty on a global scale.

In summary, food sovereignty presents a viable pathway to ensuring food security, receiving significant attention over the past two decades. Consequently, there is an increasing focus on transitioning to agroecology to implement more efficient farming methods (Peredo Parada et al., 2020). Ecologically driven management strategies in agricultural landscapes and ecosystems can enhance sustainability by reducing off-site impacts while boosting agricultural output (Matson et al., 1997).

Small-scale farming methods have shown the capacity to adopt effective techniques without depending on chemical inputs or machinery (Altieri et al., 2012). In numerous developing countries, traditional farmers have developed sophisticated agricultural methods passed down through generations. Approximately 3 million hectares of land are still managed using traditional approaches like terraces, raised fields, agroforestry systems, and polycultures, highlighting the efficacy of indigenous agricultural practices (Koohafkan & Altieri, 2011).

These traditional agricultural systems possess characteristics that enhance stability and resilience, reducing the susceptibility to losses even in adverse conditions. Moreover, they offer a diverse array of essential goods crucial for food and livelihood security (Quiroz-Guerrero et al., 2020). These systems thrive on the innovation of farmers, advancements in technology and traditional knowledge, resulting in significant biodiversity that plays a crucial role in regulating ecosystem function and providing essential ecosystem services (Singh M., 2021).

Distinctive features of traditional agroecosystems encompass innovative landscapes, efficient management of water and land resources, and conservation practices aimed at improving ecosystem management. Cultural norms, such as well-established institutions for agroecological management and customary agreements concerning resource access and sharing, are deeply ingrained in cultural values (Marchi et al., 2018; Tittonell, 2020).

Agricultural systems are influenced by a range of factors, including population growth, technological progress, market dynamics, climate change, and societal demands for land reform, food sovereignty, and poverty alleviation (Harmanny and Malek, 2019). Agroecologists have bolstered smallholder agricultural systems by implementing agroecological principles, thereby enhancing their adaptability to these factors and promoting sustainability in a continually changing world (Barrios et al., 2020). Agroecological-based systems that achieve both high productivity and resilience adhere to fundamental principles of sustainable agricultural management (Sandhu, 2021). The goals include:

1. Boosting farm productivity,
2. Strengthening resilience and managing risks,
3. Fostering biodiversity, ecological services, and preserving resources,
4. Upholding social justice, cultural diversity, and economic well-being,
5. Promoting the utilization of renewable resources and optimizing natural cycles, and
6. Preventing environmental and land degradation.

Agroecological systems prioritize the advancement of food sovereignty, advocating for universal access to safe, nutritious, and culturally appropriate food in quantities and quality sufficient to sustain a healthy life while respecting human dignity.

The agroecological approach aims to enhance energy and technological independence in anticipation of rising fuel and input costs (Ramankutty and Dowlatabadi, 2021).

Therefore, we suggest that incorporating agroecological methods into food systems necessitates a thorough evaluation of the combined effects of living organisms and abiotic factors on ecological services and agro-ecosystem functions.

Farmers have devised agroecological strategies to enhance soil quality, promote plant health, increase yields, and strengthen socio-ecological resilience, autonomy, and food sovereignty. These strategies stem from emerging practices and social movements, highlighting the importance of establishing a supportive framework for agroecological practices through responsible governance.

Additionally, the adoption of agroecological production and consumption is facilitated by favorable conditions and innovative approaches that encourage social and institutional advancement. Simultaneously, the implementation of circular and solidarity economy principles underscores the development and reinforcement of local economies by aligning solutions with local needs, resources, and capacities.

### 12.3.2. Agro-Ecological Methods Bolster Biodiversity Preservation and Advocate for Social Equity

Agricultural practices grounded in agroecological principles have long been the norm in India, spanning thousands of years. Traditional agricultural methods in India exhibit distinctive features such as intercropping, crop varietal combinations, crop rotations, weedy margins around gardens, mulching, ridging, and bush fallows. Similarly, in Uganda, coffee-banana systems are prevalent, for bananas provide shade for coffee plants (Ssebunya et al., 2019). In Nigeria and Ghana, a cocoa-plantain system is observed with developing plantains providing shade for immature cocoa trees, as detailed by Dzomeku et al., (2008) and Agbongiarhuoyi et al., (2016). Additionally, coffee cultivation alongside various tree species, such as Markhamia and Ficus, is practiced, utilizing trees as sources of lumber, fuel, and fodder. The root systems of these trees, along with associated litter, contribute to mulching, fostering a diverse community of soil microorganisms that enhance soil health (Kalanzi, 2011). Notably, mulched banana gardens in Africa are often inhabited by predatory ants, which help control pests such as banana weevils, thus contributing to pest management in plantations (Abera-Kalibata et al., 2006; Okolle et al., 2020).

Intercropping, a common practice, involves cultivating different crop varieties together, reducing pest and disease incidence while enhancing food security by increasing crop availability (Mulumba et al., 2012). Across Africa, soil and water conservation measures such as "soil basins," ridging, and the use of animal manure are widely employed. The diverse plant life resulting from agroecological methods supports a rich biodiversity, including beneficial organisms such as pollinators and natural predators, fostering soil health and overall plant vitality.

Although ecological farming can mitigate many of the negative aspects of conventional farming, it is important to acknowledge that it is typically characterized by laborious work and may not provide a definitive solution to India's food issues. The present rates of population growth and urbanization may hinder the implementation of some practices, and a significant portion of the young population may choose alternate sources of income instead of engaging in farming. The issue of decreasing soil fertility and unpredictable rainfall cannot be disregarded; yet it necessitates the use of technical innovations that can lead to substantial improvements in crop productivity.

What is necessary is a thoughtfully planned integration of the most efficient techniques offered by both traditional and ecological farming. For instance, smaller technologies like 'walking tractors' may be prioritized overusing oxen. Solar-powered water pumps for drip irrigation can serve as alternatives to rain-fed farming and large-scale irrigation projects. Motor-driven threshers and communal silos might be favored over manual threshing and traditional homestead granaries. These technological advancements can mitigate environmental impact while also decreasing the labor and monotony associated with traditional agricultural practices.

While ecological farming can alleviate many drawbacks of conventional farming, it's important to recognize its association with labor-intensive work and its potential limitations in solving the country's food challenges. The current rates of population growth and urbanization might impede the adoption of specific practices, and a significant portion of the young population may opt for alternative sources of income instead of pursuing farming.

Addressing declining soil fertility levels and unpredictable rainfall is imperative. However, embracing technological innovations with the potential to significantly boost crop productivity is crucial. Perhaps a meticulously crafted blend of the most effective methods from traditional and ecological agriculture is needed.

## 12.4. Scientific Practices and Agricultural Innovations

The following explanation makes evident that natural agricultural sciences are inextricably linked in food production systems and among the stakeholders involved, as the followingagroecological transformation presents new opportunities as well as problems for study. Agroecology asks scientists about the subjects of their studies, the techniques they create and employ, and the outcomes they yield. Rather than suggesting that biotechnology research in traditional agriculture is out of date, this section looks at more comprehensive ways that scientists might include complexity and ambiguity into their work. Ignoring these obstacles would restrict the scope of scientific inquiry and practice, which would impede the advancement of agroecological innovation (Vanloqueren et al., 2009). First, research needs to be multidisciplinary to stimulate innovation; this is consistent with agroecology's intricate and comprehensive methodology. This trend is called "interdisciplinary research," which is the practice of doing research using several academic disciplines that are unconnected to one another and have different research paradigms (Baveye et al., 2014). Agricultural, ecological, and many other fields may be connected to create novel approaches that promote restoration.

Duru et al., (2011) demonstrated the management of permanent grasslands by developing an evaluation tool in collaboration with farmers, considering the array of ecosystem services (E.S.) provided by such ecosystems. The integration of diverse knowledge systems, encompassing both scientific and practical insights, facilitates the consideration of contextual socio-ecological complexity, supporting the transition to agroecology and fostering tailored innovations.

There is ongoing debate regarding the organization of agroecology, whether as an independent discipline or as a practice that intersects or integrates various disciplines, akin to discussions in sustainability sciences. This raises the question: should diverse knowledge outputs dynamically contribute, or do we need a singular scientific discipline to address the full spectrum of sustainability challenges or agroecological issues? (Dalgaard et al., 2003). Beyond this epistemological quandary, interdisciplinary and transdisciplinary activities are deemed complementary modes of learning, as agroecology inherently relies on a diversity of knowledge sources (Chantre et al., 2014).

Despite ample evidence supporting the potential for interdisciplinary and transdisciplinary research approaches, scientists encounter challenges in translating these ideas into practice. Establishing shared objectives from the outset and addressing jointly defined problems are imperative to ensure socially resilient innovations (Méndez et al., 2013). Genuine co-leadership between science and practice is essential, with both knowledge systems being acknowledged and integrated to achieve greater symmetry (Scholz et al., 2015). The time investment in such endeavors may vary between social and natural sciences due to differing knowledge production rates. Regular feedback and dialogue among all project participants are crucial to realign research or techniques as necessary to meet the goals of both practitioners and scientists (Cuéllar-Padilla et al., 2011).

In essence, effective communication is pivotal for fostering mutual learning, establishing a trusting environment, and ensuring socially resilient outcomes (Méndez et al., 2013).

### 12.4.1. Harnessing the Capabilities of Green Technology

The stakeholder panel selected green technologies in the second and third workshops that may be able to realize the environmental visions developed in the first place. A list of new technologies and the establishment of two distinct agricultural concepts—each using a separate set of standards to determine how agriculture may become more ecologically sound—were the outcomes of these conversations.

### 12.4.2. Technology or Information and Communication (I.C.T.)

Technology encompasses decision-support tools and more efficient methods for disseminating up-to-date information on ecologically friendly agricultural practices. Additionally, over time, technology offers novel opportunities for managing, modeling, and monitoring biological environments. From a technological perspective, these opportunities include providing consumers with information about plant and animal growth to aid in selecting safe and healthy food produced in an environmentally sustainable manner—thus ensuring a seamless journey "from soil to table." Furthermore, the logical and effective handling of data derived from complex biological processes could play a pivotal role in knowledge-based food production in the future. Production data can also be utilized to establish quantitative models of biological systems and develop decision-support tools and management programs for use by both authorities and farmers.

Given the geographical separation between consumers and producers, retailing plays a critical role in conveying consumers' preferences for quality, food safety, and environmentally sustainable manufacturing through the intricate chains involved in food production.

### 12.4.3. Harnessing Technology to Make Use of Manure

Now a days, the production of domestic animals releases significant volumes of ammonia into the atmosphere. Overuse of animal dung as a fertilizer result in the leaching of phosphate and nitrate into the environment. This results in increased laughing gas production, a significant greenhouse gas, and eutrophication.

The goals of this technology are to: lessen the nutrient load in water environments, moors, common areas, forests, etc.; reduce odors associated with the storage and delivery of animal fertilizer (de-gassing); and enhance nutrient management and, consequently, the use of resources associated with the use of domestic animal fertilizer. Although it is evident that manure technology must be integrated with energy generation, more study must be done in order to make this happen. Systems that guarantee effective, logistical, and reasonable economics must also be developed to work with agriculture's decentralized structure.

### 12.4.4. Technologizing Biomass

Enhancing and creating technologies that convert biomass into energy and high-value industrial goods in an economical and effective manner is the focus of biomass technology. Although they are intended to be employed on various centralized scales, biomass technologies are advantageous to both organic and industrial agriculture.

- The technology has the potential to create a net producer of carbon-neutral energy, reduce the need for fossil fuels (thereby enhancing supply security), and minimize greenhouse gas emissions by using biomass for fuel, power, and heating.
- Additionally, the development of bio-refineries combined with biogas can offer new products and materials and enhance flexibility and competition within the agricultural sector.
- The value of agricultural goods might be raised by combining the many technologies and products that make up biomass technology. This makes it imperative to set up an interdisciplinary research project in which the future energy infrastructure is considered.

## 12.5. Sustainability Issues in Agroecology

The declines in productivity and the degradation of ecological components and their roles in agricultural landscapes have been a source of worry in the recent past. The multiple scale approach—which incorporates stakeholder wisdom, time-dependent restorative initiatives, and scientific understanding—is necessary to restore ecological functioning. In order to preserve the intricate stabilizers of ecological services, conservation efforts associated with precision farming are a suitable means of reducing the loss of natural resources in agricultural landscapes. In order to properly examine social constructs at the local to regional level, this endeavor must tackle the scientific assessment of the farm as an "ecologically sustainable unit." When assessing farms to separate out degradative elements and pinpoint environmentally responsible activities, sustainability indices might be helpful instruments. Agroecological sustainability challenges at the intersection of socio-ecological viewpoint will be examined through an analysis of a few interconnected ideas in this study.

It is anticipated that by 2050, agricultural output will have doubled, necessitating careful control of agriculture's environmental footprint. It is suggestive of the idea that agricultural systems and natural ecosystems would continue to deteriorate, leading to a loss in ecosystem services in the end. In order to reduce the strain on land resources, the anticipated doubling of agricultural activity will need the adoption of sustainable techniques and teTechnologyn place of genetically altered crops. The farming community has clearly adapted to genetically modified crops, despite persistent criticism from ecologists and environmental conservationists. The overuse of natural resources since the beginning of modern agriculture has resulted in no sustainability and the deterioration of agricultural landscapes. A prominent concern that calls for reforming agricultural policy is the long-term effects of degradative farming methods.

Agriculture may have a detrimental impact on the environment if natural resources are overused as inputs or if they are used as pollution sink.

Since these impacts are typically non-market and their costs do not factor into market pricing, they are known as negative externalities. The term "external costs" refers to the expenses incurred by the polluter when they fail to pay the full cost of their activities, which is one of the traditional reasons for market failure. There are four main characteristics of externalities in the agriculture sector:

1. The externalities' costs are frequently disregarded
2. They frequently happen gradually
3. They frequently cause harm to people whose interests are underrepresented in political or decision-making processes
4. It's not always clear who is responsible for the externality. Farmers, for instance, typically have little motivation to stop pesticides from leaking into neighboring natural areas, the atmosphere, or water bodies because the expense of remediating the environmental effects falls on society

Similarly, producers of pesticides do not bear the whole expense of their goods as they have no negative side effects. There are fewer consensuses about the economic costs of externalities in agriculture, in part due to a lack of knowledge. According to several writers, the system of economic calculations in place today significantly underestimates the worth of natural capital both now and in the future. The valuation of ecosystem services is still a contentious topic due to issues with methodology and measurement, as well as its impact on public opinion and governmental choices. The remarkable achievements of industrialized agriculture in the last few decades have obscured important adverse effects, many of which are caused by the excessive and improper use of pesticides.

### 12.5.1. Transitioning to Tailored Solutions Rather Than Dependence on Standardized Formulas

Although the word "agroecology" is now frequently used, different people have different interpretations of what it means. All too frequently, just one of agroecology's two main components—agricultural practices and the structure of the food system—is taken into consideration when it is presented. Furthermore, while ignoring the holistic approach, several research programs purport to use the notion of agroecology.

In order to attain sustainability from farm to fork, we contend in this study that agricultural practices and food system organization are inextricably linked within the framework of agroecology. Furthermore, we contend that addressing sustainability challenges requires the use of inter- and trans-disciplinary methods.

First, as demonstrated by our research, there exist ecologically based agricultural approaches that minimize the need for outside inputs and hence improve farming's environmental sustainability. Furthermore, it has been demonstrated that the many actors involved in the food system possess the power to band together to protect their operations and ensure the practices' social significance and financial sustainability. All the examples now available should not be interpreted as copy-and-paste answers, though, as there are still issues. There are no "one-size-fits-all" answers or precise formulas in agroecologyagroecology (Lyon et al., 2011). Rather, it advocates for the creation of innovations within this specific context and proposes considering the natural and socioeconomic environment in which the food is produced. It has been demonstrated that the complexity of the problems typically necessitates working at many dimensions, integrating diverse approaches, and drawing from different types of information when contextualizing innovation processes. Therefore, the study of agroecologyagroecology necessitates the participation of several academic fields and interested parties. With this kind of study, scientists must modify how they tackle the issue; the methods they employ and the scales they operate at will be determined by the issue they must solve. Likewise, farmers encountering challenges with their animals or crops must modify their approaches to fit the unique circumstances of their farming environment (Lyon et al., 2011).

### 12.5.2. Summary and Prospective Uses of Agroecology

Agroecology serves as the foundation of sustainable agriculture, widely acknowledged for its ecological benefits in preserving resources and biodiversity while simultaneously enhancing production efficiency. In addition to its ecological advantages, agroecology fosters a supportive framework through accountable governance, circular economy principles, and initiatives promoting solidarity economy, as depicted in Figure 12.1. It encompasses not just the scientific and practical facets of agriculture but also operates as a social movement rooted in ecological principles, food sovereignty, sustainability, justice, gender equality, farmer networks, resilience, resistance, and land access. Emerged in reaction to the food and financial crises of 2008, agroecology arose as a counter to the adverse impacts of capital-intensive methods utilized during the "Green Revolution."

The prevalence of agroecological practices is on the ascent owing to their inclusive approach, community engagement, and reliance on indigenous knowledge as a guiding principle. However, despite this trend, many developing nations primarily prioritize their economic and institutional interests, allocating significant resources to agro-industrial research and development while comparatively neglecting research and development efforts for agroecological and sustainable agriculture—an integral aspect of agroecology. Addressing this issue necessitates collective endeavors to establish a conducive environment favoring the adoption of agroecological agricultural methods, entailing substantial reforms in legislation, institutions, and research and development agendas.

Establishing partnerships among diverse stakeholders and organizations engaged in the agroecological movement is crucial for enabling efficient coordination of the initiative. Besides enhancing the awareness and adoption of agroecological practices, farmers will also gain from improved access to governmental support, seeds, land, and markets for their produce. Particularly in countries like India, encouraging the uptake of agroecological farming methods involves directly engaging farmers and scientists in the research agenda. This strategy promotes active participation in disseminating innovative technologies through the 'Campesino a Campesino model,' where researchers and extension specialists play pivotal roles in facilitating knowledge transfer. Such endeavors foster the development of sustainable agroecological solutions tailored to the requirements of small-scale farmers and economically marginalized non-farming communities, thus invigorating local food systems.

### 12.5.3. Summary and Discussion

Based on the data provided in the table below, it is evident that factors such as forest land, cropping intensity, and livestock intensity per net cropped area play a crucial role in fostering environmental sustainability and positively impacting our climate. Integrated crop-livestock systems, which combine crop cultivation with livestock rearing, offer numerous benefits for agricultural sustainability. One key advantage is improved nutrient cycling, where waste from livestock serves as fertilizers for crops, enhancing soil fertility and reducing the need for external inputs like synthetic fertilizers. Additionally, the integration of livestock allows for better utilization of biomass; for instance, crop residues can be fed to animals, converting them into valuable products such as meat, milk, and manure.

**Table 1.** Key Variables and their impact on the agricultural ecosystem

| Variables | Response Environment | Impact on Ecosystem |
|---|---|---|
| "Livestock density per unit of cultivated land (Head/ha)." | "Integrated crop-livestock systems improve nutrient recycling, biomass utilization, and soil carbon storage. Livestock are quantified using a standardized measure: Adult Cattle Units per hectare of cultivated land." | It has positive impact on our ecosystem |
| Variation in Rainfall | Fluctuations in rainfall patterns disrupt crop production, resulting in crop losses and income reductions for farmers, which in turn, impact agricultural sustainability. | Variation in rainfall causes vulnerable situation of agricultural productivity. |
| Forest Land | "Forests provide essential resources to farmers such as food, fodder, timber, and various other minor forest products. Moreover, they are integral in delivering ecosystem services and supporting biodiversity conservation. The calculation involved determining the percentage of forest cover within the overall geographical area." | It has a positive impact on our ecosystem. |
| Agricultural Chemical Use Intensity | The likelihood of fertilizer imbalances leading to nutritional imbalances can directly impact crop resistance against insect pests and diseases, thus restricting agricultural sustainability. A fertilizer imbalance index was devised accordingly. | It has negative impact on agro-ecosystem stability |
| Ground Water Depletion | The growing reliance on groundwater for agricultural production has led to its excessive use, depleting aquifers faster than they can replenish. Consequently, this undermines the system's resilience to droughts and other adverse weather conditions. The percentage of groundwater development serves as the indicator variable in this context. | It increases the agroecological stress |
| Fertilizer Imbalance (index) | The likelihood of fertilizer imbalances leading to nutritional imbalances can directly impact crop resistance against insect pests and diseases, thus restricting agricultural sustainability. A fertilizer imbalance index was devised accordingly. | It has a negative effect on agro ecosystem management. |
| Cropping Intensity (%) | "Heightened cropping intensity promotes diverse and integrated farming approaches, enabling better nutrient recycling, resource optimization, and improved climate regulation, thus strengthening resilience in agricultural systems. It reduces the need for clearing forests to expand cultivated land. Cropping intensity was determined by computing the percentage ratio of gross cropped area to net cropped area." | It is positively correlated to the agro ecosystem stability. |

Moreover, these systems contribute to the accumulation of soil carbon, a crucial component of soil health and climate change mitigation. Livestock grazing and manure deposition can enhance soil organic matter content, promoting carbon sequestration in agricultural soils.

To assess the intensity of livestock integration in such systems, the concept of "Adult Cattle Unit per hectare of cultivated area" is utilized. This metric standardizes the stocking density by expressing the number of livestock in terms of Adult Cattle Units (A.C.U.) relative to the cultivated land area. This allows for a consistent comparison of livestock impact across different integrated crop-livestock systems and agricultural landscapes.

Overall, integrated crop-livestock systems represent a promising approach for enhancing agricultural productivity, environmental sustainability, and resilience to climate change. By optimizing nutrient cycling, biomass utilization, and soil carbon accumulation, these systems contribute to more efficient and sustainable food production systems. However, it has also come to our attention that improper fertilizer use, excessive groundwater depletion, and the hazardous effects of chemicals in agricultural fields can disrupt agricultural sustainability. It is imperative to raise awareness about these issues at both the farmer and ministry levels. To address this, adherence to proper guidelines and monitoring of activities by officials, manufacturers, and distributors are essential. Additionally, organizing more campaigns and exhibitions aimed at educating farmers about the harmful effects of chemicals on ecology and productivity is necessary.

- *Livestock intensity per net cropped area (Number/ha):* Integrated crop-livestock systems enhance nutrient cycling, biomass utilization, and soil carbon accumulation. The measurement of livestock is standardized as Adult Cattle Units per hectare of cultivated area. It has a positive impact on our ecosystem. Variation in Rainfall Fluctuations in rainfall patterns disrupt crop production, resulting in crop losses and income reductions for farmers, impacting agricultural sustainability. Variation in rainfall causes a vulnerable situation of agrarian productivity.
- *Forest Land*: Forests offer crucial resources to farmers, including food, fodder, timber, and various minor forest products. Additionally, they play a vital role in providing ecosystem services and aiding biodiversity conservation. The proportion of forests within the total geographical area was utilized. It has a positive impact on our ecosystem.

- *Agricultural Chemical Use Intensity*: The likelihood of fertilizer imbalances leading to nutritional imbalances can directly impact crop resistance against insect pests and diseases, thus restricting agricultural sustainability. A fertilizer imbalance index was devised accordingly. It has a negative impact on agro-ecosystem stability.
- *Ground Water Depletion:* growing reliance on groundwater for agricultural production has led to its excessive use, depleting aquifers faster than they can replenish. Consequently, this undermines the system's resilience to droughts and other adverse weather conditions. The percentage of groundwater development serves as the indicator variable in this context. It increases the agroecological stress Fertilizer imbalance (index).
- The likelihood of fertilizer imbalances leading to nutritional imbalances can directly impact crop resistance against insect pests and diseases, thus restricting agricultural sustainability. A fertilizer imbalance index was devised accordingly. It has a negative effect on agro ecosystem management.
- "Cropping intensity (%) Increasing cropping intensity encourages a wider range of farming practices, aiding in nutrient recycling, optimizing resource usage, and improving climate regulation, thus enhancing resilience in agricultural production. It reduces the need for clearing forests to expand cultivated land. Cropping intensity was evaluated by determining the percentage ratio of gross cropped area to net cropped area. It shows a positive correlation with agro-ecosystem stability."

## Conclusion

Agroecology and traditional farming systems offer a holistic and sustainable approach to agriculture, integrating ecological principles with time-tested indigenous knowledge. By emphasizing biodiversity, natural resource conservation, and soil health, these systems foster resilience against climate change while ensuring food security and rural livelihoods. Traditional farming methods, such as crop rotation, mixed cropping, and organic inputs, complement agroecological practices by reducing dependency on chemical fertilizers and pesticides, thus protecting human and environmental health. These practices not only enhance soil fertility and water retention but also promote a balanced ecosystem where pests and diseases are naturally regulated. A sustainable agricultural future lies in harmonizing modern advancements with traditional wisdom. Policies should support farmers in adopting agroecological principles through training, incentives, and research. Encouraging community participation and knowledge exchange can further strengthen the adaptability and effectiveness of these systems. By embracing agroecology and traditional farming, we can move toward an agricultural model that nurtures the land, supports farming communities, and ensures long-term food sovereignty for future generations.

## References

Abd-Aziz, S., Gozan, M., Ibrahim, M. F., & Phang, L. Y. (2022). Demand and sustainability of palm oil plantation. *Biorefinery Oil Production Plants: Value-Added Products, 1*, 11–28. https://doi.org/10.1002/9783527830756.ch2.

Abera-Kalibata, A. M., Hasyim, A., Gold, C. S., & Van Driesche, R. (2006). Field surveys in Indonesia for natural enemies of the banana weevil, *Cosmopolites sordidus* (Germar). *Biological Control, 37*, 16–24. https://doi.org/10.1016/j.biocontrol.2005.11.009.

Agbongiarhuoyi, A., Ayegboyin, K., Ogunlade, M., & Orisajo, S. (2016). Farmers' use of banana instead of plantain as a shade crop in cocoa establishment: A case of Cross River state, Nigeria. *World Rural Observations, 8*, 14–22.

Altieri, M. (1995). *Agroecology: The science of sustainable agriculture* (2nd ed.). Westview Press.

Altieri, M. A. (2002). Agroecology: The science of natural resource management for poor farmers in marginal environments. *Agriculture, Ecosystems & Environment, 93*, 1–24.

Altieri, M. A., & Toledo, V. M. (2011). The agroecological revolution in Latin America. *Journal of Peasant Studies, 38*, 587–612.

Altieri, M. A. (2015). *Agroecology: Key concepts, principles, and practices*. The World Network and SOCLA.

Altieri, M. A., & Nicholls, C. I. (2017). The adaptation and mitigation potential of traditional agriculture in a changing climate. *Climatic Change, 140*, 33–45. https://doi.org/10.1007/s10584-013-0909-y.

Altieri, M. A., Nicholls, C. I., Henao, A., & Lana, M. A. (2015). Agroecology and the design of climate change-resilient farming systems. *Agronomy for Sustainable Development, 35*(3), 869–890.

Altieri, M. A., Funes-Monzote, F. R., & Petersen, P. (2012). Agroecologically efficient agricultural systems for smallholder farmers: Contributions to food sovereignty. *Agronomy for Sustainable Development, 32*(1), 1–13.

Aziz, T., Maqsood, M. A., Kanwal, S., Hussain, S., Ahmad, H. R., & Sabir, M. (2015). Fertilizers and environment: Issues and challenges. In *Crop production and global environmental issues*. Springer International Publishing.

Babbington, A. (1999). Capitals and capabilities: A framework for analyzing peasant viability, rural livelihoods, and poverty. *World Development, 27*, 2021–2044.

Barrios, E., Gemmill-Herren, B., Bicksler, A., Siliprandi, E., Brathwaite, R., Moller, S., … & van Noordwijk, M. (2020). The 10 elements of agroecology: Enabling transitions towards sustainable agriculture and food systems through visual narratives. *Ecosystems and People, 16*, 230–247.

Baveye, P. C., & Laba, M. (2014). Moving away from the geostatistical lamppost: Why, where, and how does the spatial heterogeneity of soils matter? *Ecological Modelling, 298*, 24–38. https://doi.org/10.1016/j.ecolmodel.2014.03.018.

Bhandari, G. (2014). An overview of agrochemicals and their effects on the environment in Nepal. *Applied Ecology and Environmental Sciences, 2*(2), 66–73.

Butler, S. J., Vickery, J. A., & Norris, K. (2007). Farmland biodiversity and the footprint of agriculture. *Science, 315*, 381–384.

Chantre, E., & Cardona, A. (2014). Trajectories of French field crop farmers moving toward sustainable farming practices: Change, learning, and links with the advisory services. *Agroecology and Sustainable Food Systems, 38*(5), 573–602.

Chapin, F. S., Zavaleta, E. S., Eviner, V. T., Naylor, R. L., Vitousek, P. M., Reynolds, H., Hooper, D. U., Lavorel, S., Sala, O. E., Hobbie, S. E., Mack, M. C., & Díaz, S. (2000). Consequences of changing biodiversity. *Nature, 405*, 234–242. https://doi.org/10.1038/35012241.

Cuéllar-Padilla, M., & Calle-Collado, Á. (2011). Can we find solutions with people? Participatory action research with small organic producers in Andalusia. *Journal of Rural Studies, 27*(4), 372–383.

Dalgaard, T., Hutchings, N. J., & Porter, J. R. (2003). Agroecology, scaling, and interdisciplinarity. *Agriculture, Ecosystems & Environment, 100*(1), 39–51.

De Schutter, O. (2010). Report submitted by the Special Rapporteur on the right to food. *U.N. General Assembly, Human Rights Council, 16th Session, Agenda Item 3.*

De Schutter, O. (2014). Report of the Special Rapporteur on the right to food, Olivier De Schutter. Final report: The transformative potential of the right to food. *Human Rights Council of the United Nations.*

Delcour, A., et al. (2014). État des lieux des flux céréaliers en Wallonie selon différentes filières d'utilisation. *Biotechnologie, Agronomie, Société et Environnement, 18*(2), 181–192.

Dumont, A. M., et al. (2016). Clarifying the socio-economic dimensions of agroecology: Between principles and practices. *Agroecology and Sustainable Food Systems, 40*(1), 24–47.

Duru, M., et al. (2011). Combiner des recherches en agroécologie et des dispositifs participatifs pour construire des outils d'évaluation des prairies permanentes. *Cahiers Agricultures, 20*(3), 223–234.

Dzomeku, B. M., Staver, C., Aflakpui, G., Sanogo, D., Garming, H., Ankomah, A. A., … & Blomme, G. (2008). Evaluation of the dissemination of new banana (*Musa* spp.) technologies in Central Ghana—the role of technology characteristics. In *IV International Symposium on Banana: International Conference on Banana and Plantain in Africa: Harnessing International* (pp. 735–740).

Tegtmeier, E. M., & Duffy, M. D. (2004). *International Journal of Agricultural Sustainability, 2.*

Food and Agriculture Organization (FAO). (2018). Committee on Agriculture. *26th session: Agroecology—From advocacy to action.* Rome: FAO. http://www.fao.org/3/mx456en/mx456en.pdf.

Food and Agriculture Organization (FAO). (2019). Scaling up agroecology to achieve the sustainable development goals. *Proceedings of the Second FAO International Symposium; Apr 3–5, 2018*; Rome: FAO. http://www.fao.org/3/ca3666en/ca3666en.pdf.

FAO. (2014). *Building a common vision for sustainable food and agriculture.* Food and Agriculture Organization of the United Nations. Retrieved from http://www.fao.org/3/a-i3940e.pdf.

FAO. (2015). *Proceedings of the FAO international symposium on agroecology for food security and nutrition.* Food and Agriculture Organization of the United Nations. Retrieved from http://www.fao.org/3/a-i4729e.pdf.

FAO. (2018b). *The 10 elements of agroecology: Guiding the transition to sustainable food and agricultural systems.* Food and Agriculture Organization of the United Nations. Retrieved from http://www.fao.org/3/i9037en/i9037en.pdf.

FAO. (2018c). *Catalyzing dialogue and cooperation to scale up agroecology: Outcomes of the FAO regional seminars on agroecology.* Food and Agriculture Organization of the United Nations. Retrieved from http://www.fao.org/3/I8992EN/i8992en.pdf.

FAO. (2019c). *Council: 163rd Session. The ten elements of agroecology (Document CL 163/13 Rev.1).* Food and Agriculture Organization of the United Nations.

Fernandes, L. A., & Woodhouse, P. J. (2008). Family farm sustainability in southern Brazil: An application of agroenvironmental indicators. *Ecological Economics, 66*(2), 243–257. https://doi.org/10.1016/j.ecolecon.2008.01.027.

Francis, C., et al. (2003). Agroecology: The ecology of food systems. *Journal of Sustainable Agriculture, 22*(3), 99–118.

Francis, C. A. (1986). *Multiple cropping systems.* Macmillan.

Francis, C. A., Jones, A., Crookston, K., Wittler, K., & Goodman, S. (1986). Strip cropping corn and grain legumes: A review. *American Journal of Alternative Agriculture, 1*, 159–164. https://doi.org/10.1017/S0889189300001235.

Franco, J., Borras, J., Vervest, P., Isakson, R., & Levidow, L. (2014). *Towards understanding the politics of flex crops and commodities: Implications for research and policy advocacy.* Transnational Institute.

Daily, G. (Ed.). (1997). *Nature's services: Societal dependence on natural ecosystems.* Island Press.

Conway, G. R., & Pretty, J. (1991). *Unwelcome harvest: Agriculture and pollution.* Earthscan.

Conway, G. R. (1997). *The doubly green revolution.* Penguin.

Gashu, D., Demment, M. W., & Stoecker, B. J. (2019). Challenges and opportunities to the African agriculture and food systems. *African Journal of Food, Agriculture, Nutrition, and Development, 19*, 14190–14217. https://doi.org/10.18697/ajfand.84.BLFB2000.

Glasbergen, P. (2000). The environmental cooperative: Self-governance in sustainable rural development. *Journal of Environmental Development, 9*, 240–259.

Gliessman, S. (2011). Transforming food systems to sustainability with agroecology. *Journal of Sustainable Agriculture, 35*(8), 823–825.

Hails, R. S. (2002). Assessing the risks associated with new agricultural practices. *Nature, 418*, 685–688. https://doi.org/10.1038/nature01016.

Harmanny, K. S., & Malek, Ž. (2019). Adaptations in irrigated agriculture in the Mediterranean region: An overview and spatial analysis of implemented strategies. *Regional Environmental Change, 19*, 1401–1416. https://doi.org/10.1007/s10113-019-01494-8.

Herrero, M., Estrada, G. E., Thornton, P. K., Quiros, C., Waithaka, M. M., Ruiz, R., & Hoogenboom, G. (2007). IMPACT: Generic household-level databases and diagnosis tools for integrated crop-livestock systems analysis. *Agricultural Systems, 92*, 240–265. https://doi.org/10.1016/j.agsy.2006.03.008.

Huambachano, M. A. (2019). Indigenous food sovereignty. *New Zealand Journal of Ecology, 43*, 1–6. https://doi.org/10.20417/nzjecol.43.39.

Ikrang, E. G., Unwana, I. U., & Precious, O. E. (2022). The use of artificial intelligence in tractor field operations: A review. *Poljoprivredna Tehnika, 47*, 1–14. https://doi.org/10.5937/PoljTeh2204001G.

Pretty, J., & Ward, H. (2001). World development. *World Development, 29*(2), 209–227.

Pretty, J., Brett, C., Gee, D., Hine, R., Mason, C. F., Morison, J. I. L., Raven, H., Rayment, M., & van der Bijl, G. (2000). *Agricultural systems, 65*(2), 113–136.

Pretty, J. (1995). *Regenerating agriculture: Policies and practice for sustainability and self-reliance*. Earthscan.

Pretty, J. (2003). *Science, 302*, 1912–1915.

McNeely, J. A., & Scherr, S. J. (2003). *Ecoagriculture*. Island Press.

Pretty, J., Mason, C. F., Nedwell, D. B., & Hine, R. E. (2003). *Environmental Science and Technology, 37*(2), 201–208.

Jansen, K. (2015). The debate on food sovereignty theory: Agrarian capitalism, dispossession, and agroecology. *The Journal of Peasant Studies, 42*(1), 213–232.

Kalanzi, F. (2011). Farmers' evaluation of agroforestry tree species in Robusta coffee (*Coffea canephora* Pierre ex Froehner) cultivation systems in Bukomansimbi District, Uganda (Master's thesis). *Technische Universität Dresden, Germany*.

Kalia, A., & Gosal, S. K. (2011). Effect of pesticide application on soil microorganisms. *Archives of Agronomy and Soil Science, 57*(6), 569–596.

Koohafkan, P., & Altieri, M. A. (2011). *Globally important agricultural heritage systems: A legacy for the future*. Food and Agriculture Organization of the United Nations.

Kuyah, S., Sileshi, G. W., Nkurunziza, L., Chirinda, N., Ndayisaba, P. C., Dimobe, K., et al. (2021). Innovative agronomic practices for sustainable intensification in sub-Saharan Africa: A review. *Agronomy for Sustainable Development, 41*, 1–21. https://doi.org/10.1007/s13593-021-00673-4.

Wenhua, L. (2001). Agroecological farming systems in China. UNESCO.

Lin, B. B. (2007). Agroforestry management as an adaptive strategy against potential microclimate extremes in coffee agriculture. *Agricultural and Forest Meteorology, 144*(1–2), 85–94. https://doi.org/10.1016/j.agrformet.2006.12.00.

Liu, X., Elgowainy, A., & Wang, M. (2020). Life cycle energy use and greenhouse gas emissions of ammonia production from renewable resources and industrial by-products. *Green Chemistry, 22*, 5751–5761. https://doi.org/10.1039/D0GC02301A.

Lyon, A., et al. (2011). Farming without a recipe: Wisconsin graziers and new directions for agricultural science. *Journal of Rural Studies, 27*(4), 384–393.

Altieri, M. A. (1995). *Agroecology: The science of sustainable agriculture*. Westview Press.

Marchi, M., Ferrara, C., Biasi, R., Salvia, R., & Salvati, L. (2018). Agro-forest management and soil degradation in Mediterranean environments: Towards a strategy for sustainable land use in vineyard and olive cropland. *Sustainability, 10*, 2565. https://doi.org/10.3390/su10072565.

Marsden, T., & Murdoch, J. (2006). *Between the local and the global: Confronting complexity in the contemporary agri-food sector*. Elsevier JAI.

Matson, P. A., Parton, W. J., Power, A. G., & Swift, M. J. (1997). Agricultural intensification and ecosystem properties. *Science, 277*, 504–509. https://doi.org/10.1126/science.277.5325.504.

Méndez, V. E., Bacon, C. M., & Cohen, R. (2013). Agroecology as a transdisciplinary, participatory, and action-oriented approach. *Agroecology and Sustainable Food Systems, 37*(1), 3–18.

Morgera, E., Caro, C. B., & Durán-Marín, G. (2012). Organic agriculture and the law. FAO Legislative Study 107. Retrieved from http://www.fao.org/docrep/016/i2718e/i2718e.pdf.

Mulumba, J., Nankya, R., Adokorach, J., Kiwuka, C., Fadda, C., De Santis, P., et al. (2012). A risk-minimizing argument for traditional crop varietal diversity use to reduce pest and disease damage in agricultural ecosystems of Uganda. *Agricultural Ecosystems & Environment, 157*, 70–86. https://doi.org/10.1016/j.agee.2012.02.012.

Hanley, N., MacMillan, D., Wright, R. E., Bullock, C., Simpson, I., Parrison, D., & Crabtree, R. (1998). Environmental economics and management. *Journal of Agricultural Economics, 49*(1), 1–15.

Uphoff, N. (Ed.). (2002). *Agroecological innovations*. Earthscan.

Uphoff, N. (1998). Social capital: A multiperspective approach. In P. Dasgupta & I. Serageldin (Eds.), *Social capital: A multiperspective approach* (pp. 215–250). World Bank.

Röling, N. G., & Ewagemakers, M. A. (Eds.). (1997). *Facilitating sustainable agriculture*. Cambridge University Press.

Natarajan, M., & Willey, R. W. (1996). The effects of water stress on yields and advantages of intercropping systems. *Field Crops Research, 13*, 117–131. https://doi.org/10.1016/0378-4290(86)90015-8.

National Research Council. (2000). *Our common journey*. National Academy Press.

Odusola, A. (2021). Agriculture as the fulcrum of inclusive development in Africa. In *Africa's agricultural renaissance* (pp. 1–20). Palgrave Macmillan.

Okolle, N., Ngosong, C., Nanganoa, L., & Dopgima, L. (2020). Alternatives to synthetic pesticides for the management of the banana borer weevil (Cosmopolites sordidus). *CABI Reviews, 2020*, 26. https://doi.org/10.1079/PAVSNNR202015026.

Peredo Parada, S., Barrera, C., Burbi, S., & Rocha, D. (2020). Agroforestry in the Andean Araucanía: An experience of agroecological transition with women from Cherquén in southern Chile. *Sustainability, 12*, 10401. https://doi.org/10.3390/su122410401.

Pfiffner, L., et al. (2009). Impact of wildflower strips on biological control of cabbage lepidopterans. *Agricultural Ecosystems & Environment, 129*, 310–314.

Porkka, M., Guillaume, J. H., Siebert, S., Schaphoff, S., & Kummu, M. (2017). The use of food imports to overcome local limits to growth. *Earth's Future, 5*, 393–407. https://doi.org/10.1002/2016EF0004.

Quiroz-Guerrero, I., Pérez-Vazquez, A., Landeros-Sánchez, C., Gallardo-López, F., Velasco-Velasco, J., & Benítez-Badillo, G. (2020). Resilience as an adaptation strategy of agroecosystems in the light of climate change. *AGROProductividad, 13*, 61–67. https://doi.org/10.32854/agrop.v13i11.1815.

Carson, R. T. (2000). Environmental science and technology. *Environmental Science and Technology, 34*, 1413–1418.

Rajbhandari, B. P. (2017). *Bio-intensive farming system and sustainable livelihoods*. HICAST Publications.

Ramankutty, N., & Dowlatabadi, H. (2021). Beyond productivism versus agroecology: Lessons for sustainable food systems from Lovins' soft path energy policies. *Environmental Research Letters, 16*, 91003. https://doi.org/10.1088/1748-9326/ac1e3f.

Raynolds, L. T. (2004). The globalization of organic agro-food networks. *World Development, 32*(5), 725–743.

Reid, K., Flowers, P., & Larkin, M. (2005). Exploring lived experience: An introduction to interpretative phenomenological analysis. *The Psychologist, 18*, 20–23.

Rosset, P. M., & Martínez-Torres, M. E. (2012). Rural social movements and agroecology: Context, theory, and process. *Ecology and Society, 17*(3), 17.

S. Georgiou, Langford, I. H., Bateman, I. J., & Turner, R. K. (1998). Environment and planning. *Environment and Planning*, 30(4), 577–594.

Sandhu, H. (2021). Bottom-up transformation of agriculture and food systems. *Sustainability, 13*(4), 2171. https://doi.org/10.3390/su13042171.

Scholz, R. W., & Steiner, G. (2015). Transdisciplinary at the crossroads. *Sustainability Science, 10*(4), 521–526.

Simon, X., Montero, M., & Bermudez, Ó. (2020). Advancing food security through agroecological technologies: The implementation of the bio-intensive method in the dry corridor of Nicaragua. *Sustainability, 12*(3), 844. https://doi.org/10.3390/su12030844.

Singh, M. (2021). Organic farming for sustainable agriculture. *Indian Journal of Organic Farming, 1*, 1–8. https://doi.org/10.22271/ed.book.1348.

Smith, V. H., & Glauber, J. W. (2020). Trade, policy, and food security. *Agricultural Economics, 51*, 159–171. https://doi.org/10.1111/agec.12547.

Ssebunya, B. R., Schader, C., Baumgart, L., Landert, J., Altenbuchner, C., Schmid, E., et al. (2019). Sustainability performance of certified and non-certified smallholder coffee farms in Uganda. *Ecological Economics, 156*, 35–47. https://doi.org/10.1016/j.ecolecon.2018.09.004.

Swete-Kelly, D. (1996). Development and evaluation of sustainable production systems for steeplands: Lessons from the South Pacific. In *Sustainable Land Management in the South Pacific* (Network Document No 19, IBSRAM, Bangkok).

Swiderska, K. (2011). The role of traditional knowledge and crop varieties in adaptation to climate change and food security in S. W. China, Bolivian Andes, and coastal Kenya. IIED, London.

Tamburini, G., et al. (2016). Conservation tillage mitigates the negative effect of landscape simplification on biological control. *Journal of Applied Ecology, 53*(1), 233–241.

Tilman, D., Fargione, J., Wolff, B., D'Antonio, C., Dobson, A., Howarth, R., Schindler, D., Schlesinger, W. H., Simberloff, D., & Swachhamer, D. (2001). Forecasting agriculturally driven global environmental change. *Science, 292*, 281–284. https://doi.org/10.1126/science.1057544.

Tittonell, P. (2020). Assessing resilience and adaptability in agroecological transitions. *Agricultural Systems, 184*, 102862. https://doi.org/10.1016/j.agsy.2020.102862.

Tittonell, P., Gérard, B., & Erenstein, O. (2015). Trade-offs around crop residue biomass in smallholder crop-livestock systems: What's next? *Agricultural Systems, 134*, 119–128.

Tripathi, A. D., Mishra, R., Maurya, K. K., Singh, R. B., & Wilson, D. W. (2019). Estimates for world population and global food availability for global health. In R. B. Singh, R. R. Watson, & T. Takahashi (Eds.), *The role of functional food security in global health* (pp. 3–24). Academic Press.

Vandermeer, J. (1989). *The ecology of intercropping*. Cambridge University Press.

Vanloqueren, G., & Baret, P. V. (2009). How agricultural research systems shape a technological regime that develops genetic engineering but locks out agroecological innovations. *Research Policy, 38*(6), 971–983.

Werf, H. M. G., & Petit, J. (2002). Evaluation of the environmental impact of agriculture at the farm level: A comparison and analysis of 12 indicator-based methods. *Agriculture, Ecosystems & Environment, 93*, 131–145. https://doi.org/10.1016/S0167-8809(01)00354-1.

Wezel, A., Casagrande, M., Celette, F., Vian, J. F., Ferrer, A., & Peigné, J. (2014). Agroecological practices for sustainable agriculture: A review. *Agronomy for Sustainable Development, 34*(1), 1–20. https://doi.org/10.1007/s13593-013-0180-7

Yunlong, C., & Smith, B. (1994). Sustainability in agriculture: A general review. *Agriculture, Ecosystems & Environment, 49*, 99–307. https://doi.org/10.1016/0167-8809(94)90059-0.

Zhang, W., et al. (2007). Ecosystem services and dis-services to agriculture. *Ecological Economics, 64*, 253–260.

Zhu, Y., Fen, H., Wang, Y., Li, Y., Chen, J., Hu, L., & Mundt, C. C. (2000). Genetic diversity and disease control in rice. *Nature, 406*, 718–772. https://doi.org/10.1038/35021046.

# About the Editors

***Dr. R. K. Doharey***, a distinguished academician and researcher, held a Master of Science (Agriculture), NET, Ph.D., FSSDAT, FSRDA, and FISEE. He is currently the Professor and Head of the Department of Extension Education at Acharya Narendra Deva University of Agriculture and Technology, Kumarganj, Ayodhya, Uttar Pradesh, India. Concurrently, he serves as a member of the Board of Management at Mahatma Gandhi University of Horticulture & Forestry, Durg, Chhattisgarh, India.

With a professional journey spanning over 24 years, Dr. Doharey has made significant contributions to the field of agriculture and extension education. His academic tenure includes a notable 13-year period as a professor. His scholarly contributions encompass 140 research papers, 16 authored books, and 15 book chapters, reflecting his extensive knowledge and dedication to his field. His excellence has been recognized with 35 awards, further underscoring his significant influence in the field. Dr. Doharey's international engagements extend to 18 countries, amplifying his reach and influence. His previous roles include Assistant Dean at the College of Agriculture, Azamgarh Campus (ANDUAT, Ayodhya, U.P., India), Subject Matter Specialist in Agricultural Extension at C.S.A. University of Agriculture & Technology, Kanpur, U.P., India, and Science Writer at Uttar Pradesh Council of Agricultural Research, Lucknow, U.P., India.

***Dr. Satyapriya***, Head and Principal Scientist at the Division of Agricultural Extension, ICAR-Indian Agricultural Research Institute, New Delhi, India. His expertise spans research, extension, and teaching, focusing on climate change and Agri-Nutrition Education. Dr. Satyapriya authored 100+ research articles, showcasing a profound contribution. Beyond academia, he guides MSc and PhD students, serving as the School of Social Science coordinator at IARI, actively shaping academic initiatives. His leadership extends to successful national and international research projects in agricultural extension, solidifying his distinguished professional status. Not merely confined to academia, his passion, leadership, and commitment position him as a revered figure shaping a sustainable agricultural future.

***Mr. Aman Verma***, is a dedicated researcher in Agricultural Extension Education, currently pursuing his Ph.D. at Acharya Narendra Deva University of Agriculture & Technology, Ayodhya, India (NAAC 'A++'). He holds a Master of Science (M.Sc.) in Agricultural Extension and a Bachelor of Science

(B.Sc.) (Hons.) in Agriculture from Kulbhaskar Ashram PG College, Prayagraj (NAAC 'A'), affiliated with Prof. Rajendra Singh Rajju Bhaiya University, Prayagraj, India. Additionally, he is pursuing a Post Graduate Diploma in Rural Development (PGDRD) from Indira Gandhi National Open University (IGNOU), New Delhi (NAAC 'A++'), and a Post Graduate Diploma in Agricultural Extension Management (PGDAEM) from the National Institute of Agricultural Extension Management (MANAGE), Hyderabad, An Autonomous Organization of Ministry of Agriculture, Government of India.

Mr. Verma has made significant strides in his field. His academic achievements are complemented by his extensive career as an author and researcher. He is also an Indian Council of Social Science Research (ICSSR) Doctoral Fellowship Awardee (2024) for his doctoral research work. He has qualified the UGC-NET and has published 3 books, 2 as an associate editor and 1 as a textbook. In addition to his books, he has contributed to 20 book chapters, numerous research papers, review papers, and over 40 articles in popular magazines. He is also serving as a potential reviewer in many national and international journals. His expertise extends beyond writing, as he has also delivered radio talks and guest lectures. His collaborations with renowned publishers such as Nova Science Publishers (USA), Scrivener Publishers (UK), IGI Global Publishers (USA), Elsevier, Springer and Apple Academic Press (AAP) (USA) further highlight his influence in bridging the gap between academic research and practical applications. He actively contributes to his field through participation in training programs, workshops, national and international seminars, and conferences. Mr. Verma's outstanding contributions have been recognised with prestigious awards from various societies.

***Ms. Amrit Warshini***, born in Patna, Bihar, India is a Ph.D. Scholar in the Department of Extension Education, Acharya Narendra Deva University of Agriculture and Technology, Ayodhya, U.P., India (NAAC A++). She completed her postgraduate studies at Jawaharlal Nehru Krishi Vishwa Vidyalaya, Jabalpur, Madhya Pradesh, India and B.Sc. (Hons.) in Agriculture from Sam Higginbottom University of Agriculture, Technology and Sciences, Prayagraj, Uttar Pradesh, India. She qualified for the ICAR SRF-2023 and ICAR NET-2023 with distinction; her NET qualification and prestigious ICSSR scholarship demonstrate a commitment to academic excellence, while her advocacy for gender and social awareness infuses with her purpose. She has published many book chapters and articles and participated in seminars, trainings, National and International Conferences. She has also delivered radio talks and guest lectures. She truly believes in sharing her knowledge and experiences gained to help and pave the way for others.

# Contributors

**Aman Verma**
PhD Scholar
Department of Extension Education,
Acharya Narendra Deva University of Agriculture & Technology,
Kumarganj, Ayodhya, U.P., India

**R. K. Doharey**
Professor and Head of Department
Department of Extension Education,
Acharya Narendra Deva University of Agriculture & Technology,
Kumarganj, Ayodhya, U.P., India

**Ramesh Chand Bunkar**
PhD Scholar
Agricultural Extension,
ICAR- National Dairy Research Institute, Haryana, India

**Dr. Gaurav Kumar**
Assistant Professor
Department of Extension Education,
Manglayatan University, Jabalpur, Madhya Pradesh, India

**Suresh Kumar Bishnoi**
MSc. Scholar
Agricultural Extension,
ICAR-Indian Agricultural Research Insititute, New Delhi – 110012

**Yash Pateriya**
PhD Scholar
Department of Extension Education,
Acharya Narendra Deva University of Agriculture & Technology,
Kumarganj, Ayodhya, U.P., India

**Alaka S Balan**
Assistant Professor
Department of Agricultural Extension
College of Agriculture, Ambalavayal, Kerala Agricultural University,
Kerala, India

**Sreelekshmy Sadanand**
Research Scholar
Department of Agricultural Extension
College of Agriculture, Punjab Agricultural University, Ludhiana,
Punjab, India

**Dr. Sreeram V**
Assistant Professor
Department of Agricultural Extension
Regional Agricultural Research Station, Ambalavayal, Kerala
Agricultural University, Kerala, India

**Dr. Navitha Raj**
Assistant Professor
Department of Agricultural Extension
College of Agriculture, Ambalavayal, Kerala Agricultural University,
Kerala India

**Shivam Singh**
PhD Scholar
Department of Extension Education,
Chandra Sekhar Azad University of Agriculture & Technology, Kanpur,
U.P., India

**Dr. Harsish Chandra Singh**
Professor and Head of Department
Department of Extension Education,
Chandra Sekhar Azad University of Agriculture & Technology, Kanpur, U.P., India

**Pawan Gupta**
PhD Scholar
Department of Extension Education,
Chandra Sekhar Azad University of Agriculture & Technology, Kanpur, U.P., India

**Kapil Verma**
MSc. Scholar
Department of Extension Education,
Acharya Narendra Deva University of Agriculture & Technology, Kumarganj, Ayodhya, U.P., India

**Amrit Warshini**
PhD Scholar
Department of Extension Education,
Acharya Narendra Deva University of Agriculture & Technology, Kumarganj, Ayodhya, U.P., India

**Dr. Vineeta Kumari**
Deputy Director General, Gender Studies
National Institute of Agricultural Extension Management, Hyderabada, Telangana India

**Dr. Sugandha Munshi**
Senior Scientist
International Rice Research Institute-South Asia Regional Center, Varanasi, Uttar Pradesh, India

**Vineeta Chandra**
PhD Scholar
Department of Extension Education,
Centurian University, Orrisa, India

**Anurag Dixit**
PhD Scholar
Department of Extension Education,
Acharya Narendra Deva University of Agriculture & Technology,
Kumarganj, Ayodhya, U.P., India

**Dr. Satyapriya**
Head and Principal Scientist
Agricultural Extension,
ICAR-Indian Agricultural Research Institute, New Delhi – 110012

**Venu Lenin**
Principal Scientist
Agricultural Extension,
ICAR-Indian Agricultural Research Institute, New Delhi – 110012

**Settipalli Sravani**
PhD Scholar
Agricultural Extension,
ICAR-Indian Agricultural Research Institute, New Delhi – 110012

**Mathi Girishma**
MSc. Scholar
Agricultural Extension,
ICAR-Indian Agricultural Research Insititute, New Delhi – 110012

**M. Vamshi**
PhD Research Scholar
Dairy Extension Division
ICAR-National Dairy Research Institute, Haryana, India

**Siddesh Sunil Zade**
PhD Research Scholar
Dairy Extension Division
ICAR-National Dairy Research Institute, Haryana, India

**Apoorva Veldandi**
PhD Research Scholar
Dairy Extension Division
ICAR-National Dairy Research Institute, Haryana, India

**Dr. Sanjit Maiti**
Senior Scientist
Dairy Extension Division
ICAR-National Dairy Research Institute, Haryana, India

**Dr. Sanchita Garai**
Senior Scientist
Dairy Extension Division
ICAR-National Dairy Research Institute, Haryana, India

**Dr. Chethan Patil N. D.**
Assistant Professor
Department of Agricultural Economics & Extension, School of Agriculture, Lovely Professional University. Phagwara, Punjab, India

**Dr. Sapna Jarial**
Associate Professor
Department of Agricultural Economics & Extension, School of Agriculture, Lovely Professional University. Phagwara, Punjab, India

**Rahul Dundesh Bellagi**
Research Associate
Indian Council of Social Science Research - Centre for Multi-disciplinary Development Research,
Dharwad, Karnataka, India

**Dayanand Patil**
PhD Scholar
Division of Dairy Extension,
ICAR-National Dairy Research Institute, Karnal, Haryana, India

**Seema Kujur**
PhD Research Scholar
Division of Agricultural Extension,
ICAR-Indian Agricultural Research Institute, New Delhi,, India

**Dr. Subhashree Sahu**
Scientist
Division of Agricultural Extension,
ICAR-Indian Agricultural Research Institute, New Delhi, India

**Dr. Manjeet Singh Nain**
Professor and Principal Scientist
Division of Agricultural Extension,
ICAR-Indian Agricultural Research Institute, New Delhi, India

**Dr. Sukanya Barua**
Scientist (Sr. Scale)
Division of Agricultural Extension,
ICAR-Indian Agricultural Research Institute, New Delhi-110 012.

**Dr. Pradyumna Tripathi**
Professor and Head of Department
Department of Vegetable Science, College of Agriculture
Odisha University of Agriculture and Technology, Bhubneshwar-751003, Odisha, India

**Shaili**
PhD Scholar
Department of Vegetable Science, College of Agriculture
Odisha University of Agriculture and Technology, Bhubneshwar-751003, Odisha, India

**Dr. Dipika Sahoo**
Associate Professor (Vegetable Science)
College of Horticulture, Chiplima, Sambhalpur, OUAT-768025

**Adupa Shanmuka**
Scientist
ICAR-Directorate of Groundnut Research, Gujrat, India

**Tummala Naga Satya Sai Srivani**
PhD Research Scholar
Junagarh Agricultural University, Gujrat, India

**Veesam Haripriya**
PhD Research Scholar
Division of Agricultural Extension,
ICAR-Indian Agricultural Research Institute, New Delhi, India

**Laksheeta Chauhan**
PhD Scholar
Department of Extension Education,
Rajasthan College of Agriculture,
MPUAT, Udaipur (Raj) 313001

**S. K. Meena**
PhD Scholar
Department of Extension Education,
Sri Karan Narendra Agriculture University, Jobner, Jaipur (Raj) 303329

**Devendra Singh**
PhD Scholar
Department of Extension Education,
Rajasthan College of Agriculture,
MPUAT, Udaipur (Raj) 313001

**Risabh Singh Gaur**
PhD Scholar
Division of Dairy Extension,
ICAR- National Dairy Research Institute,
Karnal (Haryana) 132001

**Smaranika Mohanty**
PhD Research Scholar
Department of Extension Education,
Siksha 'O' Anusandhan University, Bhubaneswar, Odisha, India

**Somya Sakti Dash**
PhD Research Scholar
Department of Extension Education,
Siksha 'O' Anusandhan University, Bhubaneswar, Odisha, India

**Professor Santosh Kumar Rout**
Professor and Dean
Department of Extension Education,
Siksha 'O' Anusandhan University, Bhubaneswar, Odisha, India

**Dr. Manoranjan Dash**
Professor
Information Systems & Operation Management,
Siksha 'O' Anusandhan University, Bhubaneswar, Odisha, India

**Dr. Anshuman Jena**
Associate Professor
Faculty of Agricultural Sciences
Siksha 'O' Anusandhan University, Bhubaneswar, Odisha, India

**Dibyajyoti Behera**
PhD Research Scholar
Department of Extension Education,
Siksha 'O' Anusandhan University, Bhubaneswar, Odisha, India

**Sweta Sahoo**
PhD Research Scholar
Department of Extension Education,
Siksha 'O' Anusandhan University, Bhubaneswar, Odisha, India

**Swatee Pragyan**
PhD Research Scholar
Department of Extension Education,
Siksha 'O' Anusandhan University, Bhubaneswar, Odisha, India

# Index

## A



## B



## C

## N



## P



## R



## S



## T

## V



## W